AIR, WATER AND SOIL POLLUTION SCIENCE
AND TECHNOLOGY SERIES

SULPHATE-REDUCING BACTERIA IN BIOLOGICAL TREATMENT WASTEWATERS

AIR, WATER AND SOIL POLLUTION SCIENCE AND TECHNOLOGY SERIES

Trends in Air Pollution Research
James, V. Livingston (Editor)
2005. ISBN: 1-59454-326-7

Agriculture and Soil Pollution: New Research
James, V. Livingston (Editor)
2005. ISBN: 1-59454-310-0

Water Pollution: New Research
A.R. Burk (Editor)
2008. ISBN: 1-59454-393-3

Air Pollution: New Research
James, V. Livingston (Editor)
2007. ISBN: 1-59454-569-3

Air Pollution Research Advances
Corin G. Bodine (Editor)
2007. ISBN: 1-60021-806-7

Marine Pollution: New Research
Tobias N. Hofer (Editor)
2008. ISBN: 978-1-60456-242-2

Complementary Approaches for Using Ecotoxicity Data in Soil Pollution Evaluation
M. D. Fernandez and J. V. Tarazona
2008. ISBN: 978-1-60692-105-0

Complementary Approaches for Using Ecotoxicity Data in Soil Pollution Evaluation
M. D. Fernandez and J. V. Tarazona
2008. ISBN: 978-1-60876-411-2 (Online Book)

Lake Pollution Research Progress
Franko R. Miranda and Luc M. Bernard (Editors)
2008. ISBN: 978-1-60692-106-7

Lake Pollution Research Progress
Franko R. Miranda and Luc M. Bernard (Editors)
2008. ISBN: 978-1-60741-905-1 (Online Book)

River Pollution Research Progress
Mattia N. Gallo and Marco H. Ferrari (Editors)
2009. ISBN: 978-1-60456-643-7

Heavy Metal Pollution
Samuel E. Brown and William C. Welton (Editors)
2008. ISBN: 978-1-60456-899-8

Heavy Metal Pollution
Samuel E. Brown and William C. Welton (Editors)
2008. ISBN: 978-1-61668-049-7 (Online book)

Cruise Ship Pollution
Oliver G. Krenshaw (Editor)
2009. ISBN: 978-1-60692-655-0

Water Purification
Nikolaj Gertsen and Linus Sønderby (Editors)
2009. ISBN: 978-1-60741-599-2

Water Purification
Nikolaj Gertsen and Linus Sønderby (Editors)
2009. ISBN: 978-1-61668-275-0 (Online book)

Environmental and Regional Air Pollution
Dean Gallo and Richard Mancini (Editors)
2009. ISBN: 978-1-60692-893-6

Environmental and Regional Air Pollution
Dean Gallo and Richard Mancini (Editors)
2009. ISBN: 978-1-60876-553-9 (Online Book)

Industrial Pollution including Oil Spills
Harry Newbury and William De Lorne (Editors)
2009. ISBN: 978-1-60456-917-9

Traffic Related Air Pollution and Internal Combustion Engines
Sergey Demidov and Jacques Bonnet (Editors)
2009. ISBN: 978-1-60741-145-1

Sludge: Types, Treatment Processes and Disposal
Richard E. Baily (Editor)
2009. ISBN: 978-1-60741-842-9

National Air Quality: Status and Trends
Joseph DiSano *(Editor)*
2010. ISBN: 978-1-60741-513-8

Poisoning and Acidification of the Earth's Oceans
Geoffrey E. Mason *(Editor)*
2010. ISBN: 978-1-60741-560-2

Sulphate-Reducing Bacteria in Biological Treatment Wastewaters
Dorota Wolicka
2010. ISBN: 978-1-60876-931-5

Air Pollution and Ship Emissions
Jacob Boutin *(Editor)*
2010. ISBN: 978-1-60876-087-9

Heavy Metal Compounds in Soil: Transformation upon Soil Pollution and Ecological Significance
Tatiana M. Minkina, Galina V. Motusova, Olga G. Nazarenko and Saglara S. Mandzhieva
2010. ISBN: 978-1-60876-466-2

From Soil Contamination to Land Restoration
Claudio Bini
2010. ISBN: 978-1-60876-853-0

Green Plants and Pollution
Rajiv Sinha
2010. ISBN: 978-1-61668-147-0

Green Plants and Pollution
Rajiv Sinha
2010. ISBN: 978-1-61668-403-7 (Online book)

AIR, WATER AND SOIL POLLUTION SCIENCE
AND TECHNOLOGY SERIES

SULPHATE-REDUCING BACTERIA IN BIOLOGICAL TREATMENT WASTEWATERS

DOROTA WOLICKA

Nova Science Publishers, Inc.
New York

Copyright © 2010 by Nova Science Publishers, Inc.

All rights reserved. No part of this book may be reproduced, stored in a retrieval system or transmitted in any form or by any means: electronic, electrostatic, magnetic, tape, mechanical photocopying, recording or otherwise without the written permission of the Publisher.

For permission to use material from this book please contact us:
Telephone 631-231-7269; Fax 631-231-8175
Web Site: http://www.novapublishers.com

NOTICE TO THE READER

The Publisher has taken reasonable care in the preparation of this book, but makes no expressed or implied warranty of any kind and assumes no responsibility for any errors or omissions. No liability is assumed for incidental or consequential damages in connection with or arising out of information contained in this book. The Publisher shall not be liable for any special, consequential, or exemplary damages resulting, in whole or in part, from the readers' use of, or reliance upon, this material.

Independent verification should be sought for any data, advice or recommendations contained in this book. In addition, no responsibility is assumed by the publisher for any injury and/or damage to persons or property arising from any methods, products, instructions, ideas or otherwise contained in this publication.

This publication is designed to provide accurate and authoritative information with regard to the subject matter covered herein. It is sold with the clear understanding that the Publisher is not engaged in rendering legal or any other professional services. If legal or any other expert assistance is required, the services of a competent person should be sought. FROM A DECLARATION OF PARTICIPANTS JOINTLY ADOPTED BY A COMMITTEE OF THE AMERICAN BAR ASSOCIATION AND A COMMITTEE OF PUBLISHERS.

LIBRARY OF CONGRESS CATALOGING-IN-PUBLICATION DATA

Available upon request.

ISBN: 978-1-60876-931-5

Published by Nova Science Publishers, Inc. ✦ New York

CONTENTS

Preface		**xi**
Chapter 1	**Introduction**	**1**
Chapter 2	**Ecophysiology of Sulphate Reducing Bacteria (SRB)**	**7**
Chapter 3	**Application of Sulphate Reducing Bacteria in Biological Treatment of Wastewaters**	**17**
Conclusion		**39**
Acknowledgments		**41**
References		**43**
Index		**51**

Preface

Human activity is strictly linked with the production of waste, i.e., materials and substances that are undesired and cannot be used further. On the one hand these substances are natural to the environment, eliminated from further technological process by their uselessness (e.g. mining waste), or represent new products such as anthropogenic waste, being the by-product of industrial and agricultural activities. A separate group comprises municipal waste that is not linked with production but results from human dwelling.

Utilization actions aiming at neutralizing and/or removal of waste are focused on substances that due to their existing or potential chemical activity may negatively influence the biosphere. Non-active substances represent alien elements in the natural environment, but due to their passive character, their utilization is concentrated on non-conflicting storage. Active pollutants influencing the natural environment penetrate it as gaseous emanations, fluids (sewage and effluents) and solids.

This book addresses this very important issue and covers the topic of restriction of emission and removal of hazardous gaseous emanations that should be conducted in places where they are formed. Imperfection of the applied technology or its lack results in atmospheric pollution. This problem can be of local (around industrial plants, e.g. chemical works, food processing plants, around farmsteads and stock farms), country or global range (emission of CO_2, nitrogen compounds, gases hazardous to the ozone layer).

Chapter 1

1. INTRODUCTION

1.1. TRENDS IN ENVIRONMENTAL BIOTECHNOLOGY

Human activity is strictly linked with the production of waste i.e. materials and substances that are undesired and cannot be used further. On the one hand these substances are natural to the environment, eliminated from further technological process by their uselessness (e.g. mining waste), or represent new products i.e. anthropogenic waste, being the by-product of industrial and agricultural activities. A separate group comprises municipal waste that is not linked with production but results from human dwelling.

Utilization actions aiming at neutralizing and/or removal of waste are focused on substances that due to their existing or potential chemical activity may negatively influence the biosphere. Non-active substances represent alien elements in the natural environment, but due to their passive character, their utilization is concentrated on non-conflicting storage. Active pollutants influencing the natural environment penetrate it as gaseous emanations, fluids (sewage and effluents) and solids.

Restriction of emission and removal of hazardous gaseous emanations should be conducted in places where they are formed. Imperfection of the applied technology or its lack results in atmospheric pollution. This problem can be of local (around industrial plants, e.g. chemical works, food processing plants, around farmsteads and stock farms), country or global range (emission of CO_2, nitrogen compounds, gases hazardous to the ozone layer). Control of such pollutants beyond their source areas is difficult or even impossible.

Control of liquid pollutants such as sewage or effluents or solid waste is always focused on their chemical transformation in order to obtain end-products that are neutral to the environment.

These methods are applied to sewage, whereas in the case of solid waste they may be used in the soil-water environment with essential and stable water supply, which as a solvent mobilizes solid compounds susceptible to leaching forming effluents within the dump sites or in their foreland, and facilitates the growth of microorganisms at the boundary between the solid and liquid phases.

The main aim of each sewage treatment method is protection of the natural environment against unfavourable influence caused by introduction of such wastes. For many years attention was drawn on disturbance of the oxygen balance caused by presence of organic and ammonium compounds. Due to this fact, pollution treatment methods were dominated by methods ensuring distinct reduction of BOD_5, COD, nitrification of ammonium compounds, and effective utilization of active sludge. Next, focus was drawn on eliminating inorganic compounds of phosphorus and nitrogen, degradation of non-biodegradable or poorly biodegradable compounds. The primary aim became, however, decreasing treatment costs, what is economically justified.

These aims can be realized using many methods and on every stage of liquid waste utilization. New equipment or technology may be introduced, or those previously applied can be modified and optimized. It should be remembered that preventing environmental pollution does not begin at the stage of sewage treatment, but much earlier, and requires wide-range activities. The most correct attitude is preventing hazards at their source, particularly in the case of industrial waste. In the first place activities should be undertaken to diminish the volume and harmfulness of pollutants that are by-products of industrial processes or steps should be taken to work out a technology treating several types of sewage and/or waste in one process. The scale of this problem can be illustrated by the dairy industry, where the volume of sewage flowing out of a dairy plant reaches several cubic meters per 24 h and comprises 0.5 to 3 times the volume of processed milk. Similar quantities of sewage are produced by oil refineries, where 1 ton of processed oil results in about 10 to 18 cubic meters of refinery-petrochemical waste. Analogous results have been observed in the case of solid waste produced during technological processes, where the volume of waste distinctly exceeds the final product. For example, the industry of phosphorus fertilizers produces 5 t of waste (phosphogypsum) from 1 t of phosphorites. Such waste not only poses serious hazard to the biological equilibrium in the environment, but also distinctly increases the cost of the technological process, which should in this case also include utilization of liquid and solid waste.

Treatment of different types of sewage requires application of many physical, chemical and mechanical methods. They include e.g. retaining of suspensions from sewage on screens, sieves and in settling tanks, neutralization of acidic or alkaline sewage, adsorption of sewage components on relevant adsorbents, coagulation of non-subsiding suspensions, or extraction of sewage components by relevant solvents. Although commonly applied, these methods are not environment friendly, do not entirely solve the problem of waste neutralization, and often change only the physical-chemical composition or form of the waste. In such cases, anthropogenic waste is produced, which after many years of active production in the industrial plant may even become anthropogenic deposits. They are often the source of many rare elements e.g. elements from the lanthanoid or actinoid series (rare earths), which occur in considerable quantities in the waste, but cannot be recovered due to lack of efficient process.

Thus it seems crucial to search for pro-ecological methods focused on the utilization of industrial sewage and waste that are hazardous to the natural environment. To such methods belong biological methods that are used in a wide range in the case of some types of organic sewage. All these methods use microorganisms to remove organic and some inorganic compounds from sewage. They decrease the volume of pollutants in the sewage and retain correct parameters determined by norms for sewage water introduced into surface waters. Selection of the appropriate method of sewage treatment depends on the type of sewage, its composition, volume, as well as degree of pollution of the water reservoir, to which the refined sewage will be introduced. These methods can be variously modified, and in some cases multistage sewage treatment is carried out with application of different methods.

Biological methods may be applied only to the treatment of sewage, in which the concentration of toxic compounds does not hamper the incubation of microorganisms. Due to this fact, continuous cultures are applied in biological treatment plants, which allow obtaining the maximal incubation speed of microorganisms through continuous supply of fresh medium and removal of metabolism products. In a classical continuous culture (chemostat), the time of sewage flow should correspond to the microorganism growth speed, which is determined experimentally by change of flow speed from the moment when the biomass stable level is attained.

Recently, search is focused on methods that would allow simultaneous biodegradation of several wastes. Concurrent biodegradation of two industrial wastes seems an interesting issue from the economy of the process. Costs linked with simultaneous biodegradation of two hazardous industrial wastes are always lower than for each of them separately. Biotransformation of phosphogypsum in the environment of organic sewage as a liquid state to its dissolution may be an interesting example. This trend in environmental biotechnology results in a large number of reports focused on treatment of high-sulphur sewage using sulphidogenesis methods (Lens et al., 1998) or biotransformation of solid waste in organic sewage environments. There are also single publications on the application of sulphidogenesis on the treatment of sewage after enrichment with phosphogypsum formed as a by-product in many industrial branches (Deswaef et al., 1996; Kaufman et al., 1996; Wolicka et al., 2005; Wolicka and Kowalski, 2005; Wolicka and Kowalski 2006a; Wolicka and Kowalski, 2006b; Wolicka, 2008b; Wolicka and Borkowski, 2008; Wolicka and Borkowski, 2009).

Concluding, the most aggressive in the natural environment are liquid substances and they pose the most serious hazard to the natural environment.

1.2. Advantages and Disadvantages of Anaerobic Processes

Biological treatment of various anthropogenic waste commonly applies aerobic methods, mainly due to the fast rate of the processes. However, the composition of various wastes often implies the application of anaerobic processes in the treatment procedure. In comparison to aerobic methods, this application has many advantages. First of all, anaerobic methods do not require expensive aeration, what often constitutes high costs for the whole processing plant. Further, it is estimated that only 6% COD are transferred into excessive sludge, which in many processing plants using activated sludge methodology generates additional waste. Very commonly the chemical composition of the excessive sludge does not allow it to be used e.g. in agriculture, but it can be utilized in an additional process in anaerobic conditions.

Nowadays, sulphate reducing bacteria (SRB) are becoming more frequently applied in the biodegradation of anthropogenic waste. These bacteria in course of anaerobic respiration produce hydrogen sulphide, which can bind heavy metals in poorly soluble and non-toxic sulphides of metals. This is one of the many advantages of sulphidogenesis application in environmental biotechnology. Additionally, due to the toxic activity of hydrogen sulphide, SRB may eliminate various microorganisms from the environment, including pathogenic forms, what causes their domination in a sulphate rich environment. The bacteria may be utilized during the biodegradation of two industrial wastes, of which one may be solid waste as the sulphate source, and the second – liquid waste as the carbon source. Application of such process allows simultaneous biodegradation of two arduous industrial wastes, thus shortening the biodegradation time. Moreover, post-culture deposits generated in this process i.e. carbonates and/or calcium phosphates can potentially be utilized in agriculture. An ideal example is the biotransformation of phosphogypsum.

It should be remembered, however, that anaerobic methods are not devoid of disadvantages, including:

1. difficulties in retaining the concentration of a particular microorganism group in the bioreactor due to the easy formation of symbiotic relationships, e.g. SRB easily form microbiological consortia with metanogenic archea;
2. anaerobic processes are very sensitive to pH and temperature changes and oscillation of hydraulic and substrate loading;
3. due to the multistage biodegradation process in anaerobic conditions, the duration of complete mineralization of organic compounds is much longer than in the case of aerobic processes;
4. the distinct disadvantage of common SRB application is toxic hydrogen sulphide produced during metabolic processes. This fact obliges to introduce an additional stage, during which hydrogen sulphide is oxidized e.g. to elemental sulphur. Additionally, bacteria of the genus *Desulfovibrio* influence the biocorrosion process, which is very often the case of destroyed metal elements of the hydraulic installation. This is linked with the presence of sulphates and SRB activity, what results in the reduction of oxygenated sulphur compounds, and the formation of sulphides according to the following reaction:

$$4Fe + SO_4^{2-} + 2H_2O + 2H^+ \rightarrow FeS\downarrow + 3Fe(OH)_2$$

Other metals that are present in the alloy e.g. Cu, Zn, Ni or Cr may also take part in the reaction.

Chapter 2

2. ECOPHYSIOLOGY OF SULPHATE REDUCING BACTERIA (SRB)

2.1. ENVIRONMENTS OF OCCURRENCE

Sulphate reducing bacteria (SRB) are heterotrophs and absolute anaerobes. They utilize sulphates, as well as other partly oxidized sulphur compounds (sulphites, tiosulphites and tetrationates), and elemental sulphur as the final electron acceptor in the respiration processes (Postgate, 1984; Gibson, 1990). Electron donors for this microorganism group are organic compounds such as e.g. alcohols, carboxylates, phenols, aliphatic and aromatic hydrocarbons, amino acids and some carbohydrates.

Diverse SRB physiology influences their distribution in the natural environment as well as in anthropogenic environs e.g. polluted by crude oil and oil products (Wolicka and Borkowski, 2007a; Wolicka, 2008a). The presence of SRB has been noted in aquatic and terrestrial environments (Hao et al., 1996). They occur in soils, deposits of fresh water and marine reservoirs, in silts at the moutha of river deltas etc. (Chi Ming So and Young 1999), thermal springs and in geothermal regions, in crude oil, refining and petrochemical waste, natural gas intakes and on corroding steel (Hao et al., 1996). They may be found in all types of bioreactors purifying sewage in anaerobic conditions, from which they can be isolated (Przytocka-Jusiak et al., 1997; Baena et al., 1998, 1999, 2000; Hernandez et al., 2000). The most characteristic environments of SRB occurrence are marine deposits (Bak and Widdel, 1986; Szewzyk and Pfennig, 1987; Lovley et al., 1995; Aeckersberg et al., 1998; Caldwell et al., 1999; Kniemeyer et al., 2003), in which sulphate concentration reaches averagely 28mM (Wit, 1992), as well as oil fields and crude oil reservoirs (Voordouw et al., 1996; Mueller and Nielsen, 1996;

Jenneman and Gevertz, 1999; Magot et al., 2000; Wolicka, 2008a). Their presence has also been noted in environments polluted by crude oil and oil products, dairy work sewage, whey, refining-petrochemical waste and distillery decoctions (Wolicka and Kowalski, 2005; Wolicka, 2006; Wolicka, 2008b; Wolicka and Borkowski, 2009).

Additionally, SRB always accompany crude oil and were for a long time considered as microorganisms indicative of oil deposits (Postgate, 1984). Suggestions pointing to the presence of SRB in brines from oil fields come from 1926. They are the first attempt to explain the permanent presence of sulphides in crude oil reservoirs (Jenneman and Gevertz, 1999). SRB are the most commonly isolated group from oil fields and their groundwater (Rueter et al., 1994; Mueller and Nielsen, 1996; Aeckersberg et al., 1998; Wilknes et al., 2000; Magot et al., 2000; Rozanowa et al., 2001).

Although SRB are considered as absolute anaerobics, their presence has also been noted at the boundary between the oxygenated and anaerobic zone in sediments, or even within the oxygenated zones. The presence of SRB is often recognized in the environment due to the presence of the characteristic odour of hydrogen sulphide as well as due to black colouring being the effect of precipitation of poorly soluble metal sulphides (Postgate, 1984; Gibson, 1990).

2.2. METABOLIC PROCESSES CARRIED BY SRB

Sulphate reducing bacteria (SRB) belong to absolute anaerobes, utilizing oxide compounds of sulphur as the final electron acceptors. They gain energy from oxidation of easily accessible organic compounds, and the electrons detached from the substrate are transferred on the sulphate according to the formula:

$$\text{organic compounds} + SO_4^{2-} \rightarrow H_2O + CO_2 + HS^-$$

The preferred carbon sources for SRB are low-molecule organic compounds such as organic acids e.g. lactic, formic, pyruvic, and malic acids; volatile organic acids e.g. acetic acid, and alcohols e.g. ethanol, propanol, methanol, and butanol. Some SRB species are known to utilize amino acids as the sole carbon source: *Desulfovibrio aminophilus* (Baena et al., 1998), *Desulfobacterium vacuolatum* (Rees et al., 1997), and *Desulfovibrio mexicanus* (Hernandez-Eugenio et al., 2000). Some species, e.g. *Desulfotomaculum antarcticus* may utilize glucose as the sole carbon source, but this is a rare case in SRB (Fauque et al., 1991). Rather common in turn is the utilization of aromatic and aliphatic compounds (Widdel and Bak, 1992).

All organic compounds that represent the optimal carbon source for SRB are the products of fermentation formed during anaerobic biodegradation of carbohydrates, proteins and lipids (Fauque et al., 1991; Hao et al., 1996). This results from the fact that SRB do not produce hydrolytic enzymes and become involved in the anaerobic biodegradation of organic matter in the last stage. The only exception is the archeon *Archeoglobus fulgidus*, which can produce hydrolytic enzymes.

Many papers devoted to the anaerobic biodegradation of crude oil contain information of the utilization of SRB in this process, as well as in the biodegradation of oil products both in refinery-petrochemical waste and in bioremediation of soils polluted by crude oil and oil products. Biodegradation of *n*-hexane, *n*-octane and *n*-decane by various SRB has been described by Gieg and Suflita (2002). Aeckersberg et al. (1998) described two mesophilous strains Hxd3 and Pnd3 utilizing *n*-alkanes C_{12}-C_{20} and C_{14}-C_{17}, whereas So and Young (1999) described a mesophilous strain AK-01 utilizing *n*-alkanes C_{13}-C_{18}. Tribe TD3 was able to grow on media with *n*-decane and *n*-alkanes C_6-C_{16} (Reuter et al., 1994).

Utilization of aromatic compounds such as benzene, toluene and xylene in SRB cultures has been noted by Edwards and Garbic-Galic (1992), Beller et al. (1992), Edwards et al. (1992) and Ball and Reinhard (1996). Benzene, toluene, ethylbenzene and xylene (*orto-*, *meta-*, *para*-xylene) were utilized by thermophilous sulphidogenic consortia ALK-1 and LLNL-1 described by Chen and Taylor (1997). Benzene decomposition by thermophilous SRB was noted by Lovley et al. (1995), Przytocka-Jusiak et al. (1997) and Caldwell et al. (1999).

Desulfobacula toluolica - Tol2 (Rabus et al., 1993) and *Desulfobacterium cetonicum* (Harms et al., 1999) are able to utilize toluene as the sole carbon source. The PRTOL 1 tribe isolated from soil polluted by petrol used toluene, phenyl propionate, phenyl acetate, benzoaldehyde, benzoate, *p*-cresol and *p*-hydroxybenzoate as the sole carbon source (Rabus et al., 1993). Tribe EbS7 completely oxidizing ethylbenzene was described by Kniemeyer et al. (2003). Tribes oXyS1 and mXyS1 are capable to utilize *o*-xylene (2%) and *m*-xylene (2%) as the carbon source (Harms et al., 1999). Both tribes use also toluene (2%) and benzoate.

Many SRB species are known to utilize also other organic compounds such as phenol, catechol, cresol or indole. For example *Desulfobacterium phenolicum* (Bak and Widdel, 1986) is able to biodegrade phenol as the only carbon source, *Desulfobacterium catecholicum* – catechol (Szewzyk and Pfennig, 1987), *Desulfobacterium indolicum* (tribe In04) biodegrades indole (Bak and Widdel,

1986), and *Desulfobacterium cetonicum* biodegrades *m*-cresol (Galushko and Rozanova, 1991).

The mesophilous SRB species *Desulfotomaculum gibsoniae* used phenol, catechol, metylocatechol, *m*-cresole and *p*-cresole as the sole carbon source and sulphates, sulphides and tiosulphates as electron acceptors (Kuever et al., 1999). The sulphidogenic consortium comprising *Desulfovibrio desulfuricans* A, *D. desulfuricans* B., *D. gigans* and *D. vibrio* was able to biodegrade 1,3,5-trinitrobenzene, hexahydro-1,3,5-trinitro-1,3,5-triazine and -1,3,5,7-tetranitrotetrazocine (Boopathy et al., 1998), whereas tribe BRS- HHQ20 utilized 1,2,4-trihydroxybenzene (Reichenbecher and Schink, 1997).

Over 40 genera of microorganisms taking place in the dissimilation reduction of sulphates have been subdivided into two main groups depending on the product of decomposition of the organic substrate (Brock and Madigan, 2006):

I group – bacteria that have the ability to utilize fatty acids with an even carbon number: lactate, pyruvic amid, ethanol, and some fatty acids as electron donors, but they oxidize them to acetate reducing sulphate (VI) to hydrogen sulphide, thus carbon dioxide is not produced. This type of metabolism is characteristic of e.g. *Desulfomonas, Desulfovibrio, Desulfobulbu* and *Desulfomicrobium*.

II group – bacteria that completely oxidize such compounds as e.g. fumaric acid, acetate, lactate and oxaloacetic reducing sulphates (VI) to sulphides. As a result CO_2 and H_2O are produced. To this group belong e.g.: *Desulfosarcina, Desulfonema, Desulfococcus, Desulfobacterium* and *Desulfotomaculum*.

SRB acquire energy in two main processes: phosphorylation linked with transport of electrons – accompanying sulphate reduction, as well as during oxidation of organic substrates and the accompanying substrate phosphorylation (Hao et al., 1996). The assumption that phosphorylation takes place at the level of electron transport is confirmed by the presence of cytochromes and iron-sulphur proteins and by the distinct energy profit of the process. In SRB that have the ability to use lactate and/or pyruvic acid, ATP originates in oxidation phosphorylation as well as during substrate phosphorylation, i.e. oxidation of lactate through pyruvic acid to acetate, carbon dioxide and hydrogen (Brock and Madigan, 2006).

Although SRB are able to utilize a wide variety of organic compounds, their reactions with different electron donors lead to the formation of HCO_3^- and HS^- ions (Table 1) (Hao et al., 1996).

Table 1. Metabolic reaction of SRB and the content of released energy (after Hao et al., 1996)

reaction	$\Delta G_0'$ (kJ/mol)
3 lactate → 2 propionate + acetate + HCO_3^- + H^+	-165
2 lactate + SO_4^{2-} + H^+ → 2 acetate + $2CO_2$ + HS^- + $2H_2O$	-189
4 propionate + $3SO_4^{2-}$ → 4 acetate + $4HCO_3^-$ + $3HS^-$ + H^+	-151
acetate + SO_4^{2-} → $2HCO_3^-$ + HS^-	-60
acetate + 4S + $3H_2O$ → $4H^+$ + HCO_3^- + $4HS^-$ + CO_2	-24
$4H_2$ + SO_4^{2-} + CO_2 → $3H_2O$ + HS^- + HCO_3^-	-152
H_2S + SO_4^{2-} + H^+ → HS^- + $4H_2O$	-172

Factors Influencing SRB Growth

Besides easily accessible carbon sources and the presence of oxidized sulphur compounds, many factors influence the life and growth of SRB. These physical and chemical factors include: concentration of dissolved oxygen, temperature, pH, Eh, and presence of accompanying microflora. Studies by Hao et al. (1996) have shown that concentration above 1.0 mg O_2/l leading to the increase of the redox potential, in effect inhibits SRB activity. On the other hand, some tribes such as e.g. *Desulfovibrio desulfuricans, D. vulgaris, D. desulfodismutans, Desulfobacterium autotrophicum, Desulfolobus propionicus* and *Desulfococcus multivorans* may survive at oxygen concentrations below 0.5 mg/l, even utilizing it as the electron acceptor (Dilling and Cypionka, 1990). Another important factor influencing the activity of SRB is temperature. Most tribes and sulphidogenic assemblages display the optimal activity within the range 28–32°C. There are also species that survive in higher or lower temperatures; e.g. some tribes of *Desulfobacter* prefer temperature in the range 24–28°C, and *Thermodesulfobacterium commune* – about 70°C (Hao et al., 1996). The optimal growth temperature of archaea such as *Archeoglobus fulgidus* and *A. profundus* isolated from hydrothermal waters is at 80°C. It should pointed out, however, that temperature exceeding 45°C may be killing for SRB. Similarly as temperature, pH of the environment may represent a factor inhibiting SRB growth. It is commonly considered that these organisms prefer environments with a neutral reaction, whereas pH below 5.5 and above 9.0 inhibits their growth (Hao et al., 1996). On the other hand, a number of papers point to the occurrence of SRB in acidic mine waters, where pH is about 3, or in exploitation workings where this value fallows down to 2 (Wolicka and Borkowski, 2007b).

Strong inhibiting and even toxic activity of SRB has been observed in the case of H_2S as well as HS^- and S^{2-} ions. They react with cytochromes and iron of ferredoxines, as well as with other intracellular compounds containing this element, retarding the system of electron transport (Hao et al., 1996). Other known inhibitors of sulphate reduction are selenite ions (SeO_4^{2-}), antagonists of sulphate reduction, as well as molibdate ions (MoO_4^{2-}), which are structural equivalents of SO_4^{2-} ions and cause the cellular ATP content. Recently, excessive iron ions in the environment are also considered to inhibit sulphate reduction.

SRB similarly as most microorganisms are sensitive to bacteriocides and bacteriostatics (Hao et al., 1996).

Some bacteria belonging to SRB are able to utilize electron acceptors other than SO_4^{2-} i.e. SO_3^{2-} and $S_2O_3^{2-}$. The utilization of tiosulphates and sulphites as the final acceptors allows bacteria to obtain energy indispensable for growth, because they do not reduce sulphates. A representative of this group is *Desulfovibrio sulfodismutans*, which has the ability to reduce each of these compounds according to the following reactions (Hao et al., 1996):

$$S_2O_3^{2-} + H_2O \rightarrow SO_4^{2-} + HS^- + H^+ \quad \Delta G_0' = -21.9 \text{ kJ}$$

$$4SO_3^{2-} + H^+ \rightarrow 3SO_4^{2-} + HS^- \quad \Delta G_0' = -235.6 \text{ kJ}$$

In anaerobic conditions elemental sulphur may also represent the final electron acceptor. This process is known as sulphur respiration and is not as common as sulphate reduction. A representative of this group is *Desulfuromonas acetoxidans* that utilizes acetate (rarely ethanol, propanol) as the carbon source and electron donor, and conducts its complete oxidation to CO_2 according to the following reaction (Bothe and Trebst, 1981).

$$CH_3COOH + 2H_2O + 4S^0 \rightarrow 2CO_2 + 4H_2S \quad \Delta G_0' = -5.7 \text{ kcal/mol}$$

Electrons detached during oxidation of organic compounds are transported in the respiration cycle on cytochrome c_7, next on protein 4Fe-S, and then reach the final electron acceptor – elemental sulphur. The ability to reduce elemental sulphur has been noted also in other microorganisms, among which occur:

- *Sulfospirillum* – utilizing most frequently H_2 as the electron donor.

- *Desulfurella* – thermophilous bacteria utilizing acetate as the electron donor.
- *Campylobacter* – not able to reduce sulphates but reducing sulphur, sulphites, tiosulphates, nitrates as well as fumaric acid, utilizing acetate as the electron donor.
- *Pyrodictium* – a thermophilous archeon able to utilize diatomic hydrogen as the electron donor, and elemental sulphur as electron acceptor.

Particular SRB species differ in several features, including: cell shape, mobility, occurrence, preferred electron donors, complete or incomplete oxidation of organic compounds, content of GC pairs in DNA, formation of spores, presence of desulphoviridine, cytochromes, and optimal growth temperature (Gibson, 1990).

Based on rRNA analysis, SRB have been subdivided into four groups:

1. gram -, mesophilous SRB;
2. gram +, SRB generating spores;
3. thermophilous SRB;
4. thermophilous archaeabacteria reducing sulphates.

2.3. SRB Isolation and Culture Methods

Till 1984, SRB were considered to be dominated in the natural environment by other microorganisms. Therefore, SRB were commonly isolated on media containing an easily accessible carbon source, e.g. lactate, pyruvic acid or ethanol, and sodium sulphate, well soluble in water, as the electron acceptor (Postgate, 1984). However, the last twenty years have brought many reports on the fact that SRB are microorganisms that have the ability to biodegrade a wide spectrum of organic compounds. In due course, SRB began to be isolated on media containing compounds that are the main pollutants in the environment. Thus, active SRB communities are isolated from environments polluted by substances that are subject to treatment. For example, anaerobic SRB communities can be isolated from refinery-petrochemical sewage, soils polluted by oil products, or other organic sewage, and are then used in the anaerobic utilization process of these wastes.

Very soluble sulphates such as sodium sulphate were considered the optimal electron acceptors in the sulphidogenic process. There are also reports indicating the fact that poorly soluble sulphates such as bassanite ($CaSO_4 \times 0,5H_2O$), gypsum ($CaSO_4 \times 2H_2O$), anglesite ($PbSO_4$), or barite ($BaSO_4$) may also become an easily accessible electron acceptor for SRB (Karnachuk et al., 2002).

During multiplication of anaerobic microorganism communities, typically two methods of SRB selection are applied: the *"microcosms"* method and multiplication on agar medium (Figure 1). Reproduction of selected microorganism communities should take place in strictly anaerobic conditions. Therefore, liquid media are often supplemented with compounds decreasing the redox potential, e.g. cysteine, sodium tioglicolate, or sodium sulphide, as were as indicators of oxygen level such as resaurine.

The *"microcosms"* method is considered one of the best multiplication methods of anaerobic bacteria communities from the environment. This fact was observed by Lesage et al. (2000) and Wolicka (2006, 2008a). During SRB multiplication and isolation, enrichment is based on the addition of solutions of sulphates and particular organic compounds that comprise the main pollutants, e.g. casein in the case of dairy sewage and phenol in the case of refinery-petrochemical waste.

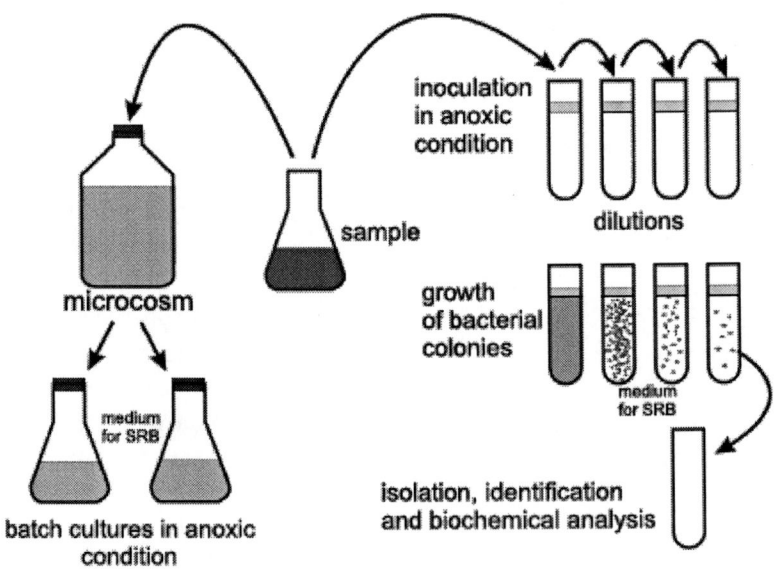

Figure 1. Methods of isolation of sulphate reducing bacteria.

The method was applied by Warren et al. (1999) during isolation of acetotrophic methanogenes; Lesage et al. (2000) during multiplication of anaerobic bacteria communities able to utilize polycyclic aromatic hydroharbons;

Chapter 3

3. APPLICATION OF SULPHATE REDUCING BACTERIA IN BIOLOGICAL TREATMENT OF WASTEWATERS

3.1. CHARACTERISTICS OF WASTEWATER IN ANAEROBIC TREATMENT

Prior to application of biological treatment methods on sewage, various physical and chemical analyses are made in order to characterize the sewage and select the appropriate treatment method. The analyses determine: pH, temperature, colour, smell, turbidity, suspension content, COD, BOD, and content of inorganic compounds. Important data include also the volume of sewage that should be treated in a given time period, as well as the sewage discharge method, i.e. whether the sewage will flow steadily or periodically to the treatment apparatus.

Sewage is produced in many branches of industry. According to the definition, sewage represents water that is polluted beyond its natural reservoirs, thus it can be assumed that precipitation water (rain, hail, snow) polluted on streets in towns and cities, as well as on the surface of fields also becomes sewage. Water drawn from surface reservoirs for consumption, sanitation or production is polluted in households and various types of service and production plants. The content and quality of sewage depends thus on the industrial development and intensity of economical activities in a given country or region, as well as on climate.

Sewage that is planned to be treated during sulphidogenesis should fulfil several prerequisites. It should contain oxygenated sulphur compounds such as e.g. sulphates that are the main electron acceptor for the SRB, as well as easily accessible organic compounds. Lactate has been previously considered as the optimal carbon source for SRB, however application of this compound as a carbon source in high-sulphur sewage is economically not feasible. Therefore the search began for other carbon sources available to SRB and constituting the main pollutants in organic sewage.

Organic sewage contains organic compounds that are either easily biodegradable by microorganisms, or, on the contrary, are poorly or very poorly biodegradable. In general, during sulphate reduction organic acids and alcohols are easily biodegradable, whereas e.g. aliphatic and aromatic hydrocarbons, pesticides, herbicides and xenobiotics are poorly biodegradable. Sewage polluted by organic compounds susceptible to biodegradation is produced mainly by food industry, e.g. dairy plants, sugar plants, fruit and vegetable processing plants, meat processing plants, breeding plants, and slaughter houses.

Dairy sewage contains high volume of organic compounds and BOD. The main constituents of this sewage are saccharides, fats and proteins that compose milk. This sewage undergoes biodegradation easily, after which the environment becomes acidified and hydrogen sulphide, toxic for most microorganisms, is produced. The sewage cannot be introduced into the receiver without treatment. The volume of sewage produced by a single dairy plant reaches several thousands of cubic meters per 24 h. The composition of dairy sewage resembles highly diluted whole milk, with dissolved lactose and protein (casein) in an identical proportion as in milk. Based on the analysis of milk composition used in the processing plant and the production profile, the sewage composition can be assumed. The volume of sewage formed in dairy plants comprises from 0.5 to 3 times the volume of processed milk. Typically, 1 g BOD_5 in sewage is considered to correspond to 9 g BOD_5 in milk. The BOD_5 of sewage reaches 450–5800 mg O_2/l (averagely about 1800 mg O_2/l).

The total volume of dairy sewage comprises strongly polluted production sewage formed during flushing of products or washing of vessels, equipment, facilities, as well as slightly polluted cooling waters (comprising 60–90% of sewage volume). Sewage may also contain low quantities of flavouring matter, gelling and clarification agents, etc., as well as disinfectants, cleaning and degreasing agents, and agents removing limescale from utensils. Sewage reaction varies between 7.0 and 8.8, with the exception of acidic sewage from production of casein and selected cheeses, which varies between 5.5 and 6.5. Due to the content of lactose, dairy sewage is easily biodegraded, and its pH falls down to very low values. The composition of sewage from particular dairy plants may vary considerably depending on the type of manufactured products (Tab. 2, 3, 4). The table 2 presents the typical composition of dairy sewage from a plant producing cream, cottage cheese and kefir.

Table 2. The typical composition of sewage from plant producing milk products

Determinations	Content [g/m³]	Determinations	Content [g/m³]
Total dry mass	1273	Potassium	41
Mineral dry mass	626	Calcium	77
Organic dry mass	647	Magnesium	15
BOD_5	587	Sodium	93
COD	1048	Sulphates	258
Total nitrogen	33	Chlorides	95
Phosphorus	13	pH	7.3

The ballast of dairy sewage is more variable when it encompasses buttermilk, whey and skimmed milk. The application of some sewage constituents e.g. whey to feed breeding animals, or buttermilk to produce beverages, may distinctly influence this ballast. Whey contains about 72% lactose, 10% protein, 0.5% fat, as well as mineral compounds and vitamins. In the dairy industry it is formed as a by-product during cheese or casein manufacture. One volume of produced cheese results in almost 10 volumes of whey, which is a crucial problem in cheese manufacture (Kutera and Talik, 1996).

Table 3. Typical composition of sewage formed during washing of pipelines for milk transport

Determinations	Content
pH	9.6
Oxidization [mg O_2 / dm^3]	290
COD [mg O_2 / dm^3]	1200
BOD_5 [mg O_2 / dm^3]	515
ether ekstract [mg / dm^3]	160
Suspensions [mg / dm^3]	480

Table 4. Typical composition of UHT milk

Determinations	Share
Lactose	4.7%
Fats	~3.2%
Proteins	~3%
Mineral salts	0.8%

Casein and lactose are not optimal carbon sources for SRB. Instead, products of their hydrolysis may become electron donors for SRB, as these bacteria do not produce hydrolytic enzymes and therefore they take part in the process of organic matter biodegradation at the level of volatile fatty acids (Mizuno et al., 1998). Due to this fact, utilization of proteins and disaccharides is a rare case in SRB.

Sewage polluted by organic compounds that are not easily biodegraded includes sewage produced in e.g. oil refinery and petrochemical plants, pulp and paper industry, textile industry, leather industry, as well as plant and animal utilization plants.

Refinery-petrochemical sewage contains from about 0.3 to 2% of organic compounds from crude oil, such as hydrocarbons, alcohols, aldehydes, esters, alkali, acids and their salts from the refining-petrochemical industry, as well as oily pollutants (with a various degree of emulsification) and tars.

In petrochemical plants sewage is formed in the process of raw material cleaning, as well as during the manufacture of many chemical products, half-products and usable products from crude oil and its fractions and from earth gas (most commonly distillation and rectification processes). Sewage composition depends on the type of production in a given petrochemical plant. The main pollutants in the sewage are hydrocarbons, alcohols, aldehydes, phenols, esters, alkali, acids and theirs salts. Other common constituents of petrochemical sewage include phenol, ethylene, propylene, butadiene, acetone, glycol, plastics, synthetic rubber, epoxy resins, and surface active agents. Petrochemical sewage is commonly treated jointly with refinery sewage.

Petroleum processing includes thermal processing comprising distillation, rectification and cracking. In modern refinery plants processes of further fuel refinement are carried out, including petrol reforming and hydrorafinery (desulphurization) of diesel oil. The first phase of crude oil processing comprises removal of gas in gas separators, mechanical pollutants by water flushing and sedimentation in settler tanks, whereas emulsions are broken in electromagnetic field, by heating or addition of deemulsifiers. After these initial stages, crude oil is distilled by separation of hydrocarbons into fractions characterized by similar boiling temperatures. At normal atmospheric pressure are produced: petrol, heavy petrol, naphtha, diesel oil and oil fuel. Distillation of fuel oil at lower pressure leads to the production of diesel oil, light, medium and heavy petrol (paraffin oils).

Hydrocarbon fractions obtained during distillation are refined in order to obtain purified products. Modern oil refinery methods are based on the adsorption of pollutants or application of selective dissolvents that dissolve the pollutants not dissolving the oil. The most commonly applied dissolvents are phenol and furfuryl alcohol.

The content of sewage produced in oil refineries depends on the quality of oil and degree of its processing, and varies from 10 to 18 cubic meters per tonne of processed crude oil.

Sewage is produced during:

- washing of facilities (high volumes of oil and oil products);
- during flushing of oil and dewatering of raw oil (fatty and naphthene acids, phenols, inorganic salts, mainly NaCl);
- during crude oil distillation and cracking (hydrogen sulphide, thiols);
- during refinery: acidic sewage (sulphuric acid, resins) and alkaline sewage (sodium hydroxide, sodium hydrosulphide, sodium salts of fatty acids, phenolates and emulsions of oil products)

The colour of sewage varies from milk white to dark brown and it has a specific odour (Table 5). The main pollutants in sewage from refining plants include naphtha, sulphides, thiols, phenols, fatty and naphthene acids, naphthene sulphoacids, mineral oils, aliphatic and aromatic hydrocarbons (often chlorinated), aldehydes and alcohols formed during oxidation of hydrocarbons, and suspensions. Naphtha occurs in sewage in forms of emulsions or surface films.

Table 5. Characteristics of sewage formed during naphtha refinery

Determinations	units	results
Temperature	°C	15–63
Odour	descriptive	naphtha and H_2S
Transparency	cm	1–4
Reaction	pH	6.5–8.4
Suspensions	mg/l	200–4200
Oxidization	mg O_2/l	650–1200
Phenols	mg/l	50–150
Hydrogen sulphide	mg H_2S/l	10–30
Oils	mg/l	20–80

Inorganic sewage that may be biodegraded during sulphidogenesis results from e.g. production of artificial fertilizers, mines, manufacture of various inorganic chemical compounds, metal ore processing plants, and electroplanting plants. Depending on the type of processing plant, it contains various types of pollutants, including biogenic elements, acids, alkali, salts and heavy metals. Toxic inorganic sewage may contain: heavy metals, inorganic acids, alkali, cyanides, etc. as the main pollutants. Such sewage is formed in metal ore and sulphur mines, in heat power plants, smelters, electroplanting plants, mineral acid works, and in explosive factories.

3.2. BIOTRANSFORMATION OF SOLID WASTE IN ORGANIC WASTEWATER

The presence of SRB has been noted in practically all reactors treating various types of organic sewage, e.g. dairy sewage (Baena, 1998, 1999, 2000; Hernandez, 2000), however they still play a small role due to the low concentration of sulphates ca. 258 g/m^3 (Talik, Kutera, 1997). The source of sulphates for SRB may be not only easily soluble compounds such as Na_2SO_4, but also insoluble mineral phases such as hanebachite ($CaSO_4$), jarosite ($KFe_3[OH]_6|(SO_4)_2$), anglesite ($PbSO_4$) or barite ($BaSO_4$) (Karnachuk et al., 2003). A good source of sulphate ions are solid anthropogenic wastes such as phosphogypsum formed during the production of phosphoric acid.

Addition of sulphates to primarily non-sulphur sewage results in high-sulphur sewage similar to sewage produced during the manufacture of molasses (2.9 g SO_4/L), citric acid from sugar cane (2.5–4.5 g SO_4/L) or wood industry (1–2 g SO_4/L) (Colleran et al., 1995). High concentration of sulphates in sewage hampers treatment using methanogenesis; in turn they can be purified with application of sulphidogenesis. In reactors treating high-sulphur sewage, SRB are solely responsible for the final stages of biodegradation of organic pollutants (Colleran et al., 1995).

Biodegradation of two different industrial wastes including sewage and solid waste seems both an interesting and indispensable procedure for economic reasons. Biotransformation of phosphogypsum in various industrial sewages such as refinery-industrial sewage, dairy sewage or distillery decoctions has already been described (Wolicka and Kowalski, 2006a; 2006b; Wolicka, 2008b). Products of phosphogypsum biotransformation and biodegradation of organic compounds in stationary cultures were carbonates and/or phosphates. The obtained results indicate the possibility of obtaining secondary post-culture deposits that can later be applied e.g. as fertilizers in farming.

Due to the lack of hydrolytic enzymes in most SRB, application of a two-step process of the treatment of organic sewage with phosphogypsum seems a good solution (Kaufman et al., 1996; Deswaef et al., 1996) (Figure 2). In the first phase, the biological reactor is colonized by an assemblage of acidogenic and sulphate reducing bacteria.

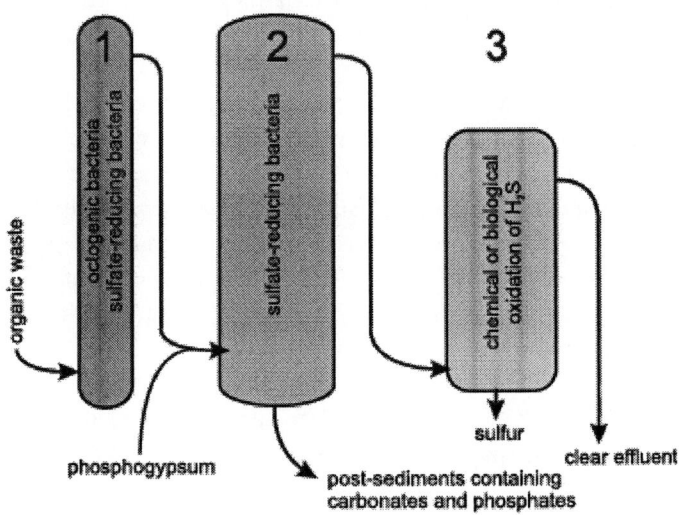

Figure 2. Treatment of organic waste water with phosphogypsum in two stages process.

Acidogenic bacteria are capable of producing acetate from organic compounds that are not easily accessible to SRB. In turn, the produced acetate can be a good energy substrate for the second group of bacteria. In the second stage, the reactor is colonized mainly by SRB that use simple organic compounds flowing from the first reactor. In both reactors waste gypsum is the source of sulphates. This method of sulphidogenesis allows for effective phosphogypsum biodegradation in the organic sewage setting, which can at first contain organic compounds that are not easily accessible for SRB (Deswaef et al., 1996). The process described herein can be conducted in bioreactors basing on the flooded deposit with biofilm. A crucial problem in these methods is the possibility of biofilm overgrowth on grains of filling material, a process that can be additionally accelerated by sparingly soluble gypsum. The possible solution may be application of reactors in which the circulating elements are the filling material grains with biofilms. As a result, excessive biofilm is removed and good contact of microorganisms with the solution of sewage with phosphogypsum is secured. Due to continuous circulation, the accumulation of non-reduced calcium sulphate in the biofilm is simultaneously hampered.

The rule of biological deposits is based on removal of pollution from sewage by microorganisms in the biofilm deposited on various types of carriers. In biological sewage treatment many types of biological deposits are applied. Selection of the deposit type depends on technological parameters of the sewage treatment plant, type of sewage and degree of sewage treatment.

The most commonly applied deposits are biological sprinkling beds comprising a container with various types of rough surface fillers e.g. pebbles, slag, plastic elements, etc. The carrier surface is covered by biofilm composed of microorganisms such as bacteria, fungi or protozoans. During the process, selection and adaptation of microorganisms to respective treatment conditions takes place. The biofilm microorganisms utilize organic and some inorganic compounds from sewage and cause decrease of their content in the sewage. Effective deposit activity depends on various parameters such as content and composition of treated sewage, temperature, pH.

Biological deposits can be used in the treatment of many different types of sewage. Characteristics of the biofilm are based on observation of microorganisms comprising it. It is also tested whether biofilm ripped off the surface of fillers does not occur in the runoff of the incubating device.

Two main types of bioreactors are currently applied in anaerobic sewage treatment:

- Reactors with biofilm on solid material (filler), such as plastic, ceramic and glass particles, sand grains, expanded clay pellets, etc. This type pf bioreactor is characterized by the fact that biomass is retained in the reactors and sewage can flow faster than it is assumed from the growth speed of microorganisms.
- Reactors without solid phase, in which granules composed of anaerobic microorganism assemblages are formed naturally. This bioreactor type increases the settling of microorganisms, what is aided by the separation of biomass from the treated sewage. Application of settling tanks in such bioreactors allows initial concentration of biomass prior to its collection, and at the same time decreases the content of microorganisms in the discharged sewage.

Application of multi-stage continuous cultures in the biotransformation of phosphogypsum seems very interesting due to the occurrence of rare earth elements. Recovery of these elements using sulphidogenesis seems purposeful due to the possibility of accumulation in the biofilm of the deposit, what has been well documented (Mueller and Steiner, 1992; Kowalski et al., 2002). The problem of accumulation of heavy metals in the secondary excessive deposit is very crucial, because on the one hand it hampers wide usage of the deposits e.g. in agriculture, but on the other hand, seems a good method of recovery of the REE scattered in phosphogypsum. It has been determined (Kowalski, 2002) that REE percentage content is higher in secondary post-culture deposits that in phosphogypsum. Due to the fact that in subsequent treatment stages the deposit mass and the content of lanthanoids decrease and assuming that the system is closed, it can be stated that the REE are retained on the biofilm (Kowalski, 2002). Problem of the concentration of scattered metals in various industrial wastes is not only the case of phosphogypsum, but is also crucial in the treatment of other solid inorganic wastes. The effect of this process are sulphides of various metals, which can be recovered also using biological methods, although in aerobic conditions, e.g. using chemolithoautotrophic sulphur bacteria.

3.3. REMOVAL OF HEAVY METALS FROM INORGANIC WASTEWATER

Heavy metals are metallic elements with density over 4.5 g/cm^3 and atomic mass exceeding 50. This group comprises such metals as: arsenic, manganese, zinc, chrome, iron, cadmium, lead, nickel, copper, mercury, cobalt, and molybdenum. Their characteristic feature is the ability of very long persistence in the natural environment, what is linked with sparing solubility of some heavy metals chemical compounds and also with large ability to accumulate in living organisms. Heavy metals can be subdivided into four groups:

- elements with very high potential hazard to environment, e.g.: cadmium, mercury, chrome, silver, zinc, gold, antimony, tin, thallium;
- elements with high potential hazard to environment, e.g.: molybdenum, manganese, iron;
- elements with medium potential hazard to environment, such as vanadium, nickel, cobalt and wolfram;
- elements with low potential hazard to environment, e.g. zirconium, tantalum, lanthanum, niobium.

Natural processes that influence the mobility of heavy metals in the environment include:

- magmatic and post-magmatic processes;
- metamorphic processes (transformation of sedimentary or magmatic rocks in high temperature or pressure, etc.);
- hypergenic processes (taking place under hydro-, atmo- and biospheric factors, e.g. weathering, erosion, transportation, sedimentation, etc.).

Due to these processes the primary occurrence of heavy metals is transformed. Migration of heavy metals as well as their uncontrolled appearance in the environment is greatly influenced by anthropogenic processes. Industrial activity, e.g. exploitation of metal ores, transport of ore to processing plant, transformation, final management and utilization of used products are the main anthropogenic sources of heavy metals. The resulting solid and gaseous wastes as well as sewage are accumulated in water, soil and atmosphere.

High content of heavy metals can be found in sewage deposits and industrial sewage. Industrial waste containing heavy metals derive mainly from: smelters, electroplating plants, tanning, fertilizer, pesticide, dyeworks, textile, and electrochemical industries, from plants producing batteries, accumulators, catalysts, etc. Copper in sewage is derived from e.g. metallurgy, dyeworks, and textile industries, and is also emitted during production of pesticides and fertilizers. Electroplating and paper industries, refineries, and steelworks supply high content of nickel to the environment, whereas production of batteries and paints, textile and plastic industries, polymer stabilizer industry, printing and graphic plants deliver high contents of zinc. Heavy metals are also introduced to water with industrial sewage and wastes, with effluents from fields or smelter dumps.

Products of sewage treatment are sewage deposits in which the heavy metal concentration distinctly exceeds that in sewage. The harmful influence of heavy metals on living organisms and the natural environment is undisputable. Therefore actions should be undertaken to eliminate or/and restrict their emission, as well as minimize their negative influence. Industrial plants in which sewage with heavy metals are produced should have installations for their preliminary treatment before sewage is sent to the receiver. Treatment of isolated sewage streams is more effective that treatment of mixed sewage.

Sewage containing heavy metals are commonly treated using: chemical (neutralization, reduction and/or oxidation, precipitation), physical and chemical (sorption, extraction, ion exchange) and electrochemical methods. The correct method depends on the type of sewage, their content, phase and concentration of particular particles and required treatment degree. Recently, biological methods are also more frequently applied (Figure 3); these methods allow recovery of heavy metals and cause the transformation of toxic cations of heavy metals into sparingly soluble sulphides, a desired effect particularly in solid waste management (Ekstrom et al., 2008; Gibert et al., 2002).

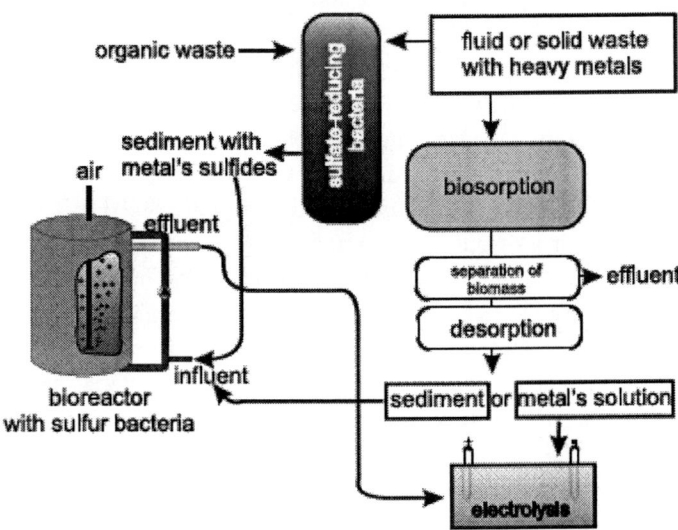

Figure 3. Biological methods to the remove heavy metals from waste water.

One of the methods of microbiological removal of heavy metals in anaerobic conditions from sewage or sewage deposits is biosorption. This process comprises: linking metal ions by reactive groups of biopolymers occurring in cell membranes of microorganisms, their preservation on surfaces of non-soluble hydroxides, salts or metal complexes, chemical reactions with released metabolites, formation of non-soluble metal compounds, followed by their accumulation and crystallization within cell membranes. Key role in this process is played by extracellular polymer substances (EPS) abundantly released by some microorganisms. Biosorption has several limitations linked with high costs caused by the need to separate biomass from the post-culture fluid after sorption and low stability of biomass, precluding multiple applications of microorganisms in subsequent cycles of sorption and desorption. Immobilization of organisms hampers their passage to the mobile phase transporting the substrate and product, allows the increase of microorganism population density in the reactor and effective separation of biomass from the solution. It also has positive influence on the stability and resistance of biomass, as well as the increase of its application in continuous processes.

Bacterial reduction of sulphates allows simultaneous removal of metal sulphates by transforming them into sparingly soluble sulphides, biodegradation of organic compounds and lower acidification of the environment. Data on the influence of heavy metals on SRB are not univocal. According to some authors, even low concentrations of heavy metals can hamper SRB activity, whereas according to others, these microorganisms have low sensitivity to the presence of heavy metals because sulphide ions produced by them during sulphate reduction allow their precipitation into non-soluble and non-toxic deposits. These discrepancies may come from the fact that the cultures are variously multiplied, i.e. at different composition, pH, and content of introduced microorganisms as well as the initial form of the introduced metal (Koschorreck, 2008). Application of SRB in bioremediation of sewage containing high concentration of heavy metals requires knowledge on the toxicity of these metals (Dvorak et al., 1992; Hass and Polpraset, 1993; Karnachuk et al., 2003). Heavy metals play also a crucial role in the metabolism of microorganisms, e.g. bacteria taking part in the regulation of biochemical processes, stabilization of cell structures or catalysis of enzymatic reactions.

SRB can be thus be utilized in treatment of sewage and mine water from toxic metals, and the microorganisms applied in this process should effectively reduce sulphates, tolerate changes in pH and be resistant against toxic metals.

Microorganisms conducting dissimilative sulphate reduction contribute to the treatment of mine and metallurgy sewage, as well as effluents from municipal and industrial waste dumps. Application of SRB in the demobilization of various heavy metals has been described in a number of papers, e.g. the influence of chromium, nickel, manganese, copper, and zinc on *Desulfovibrio vulgaris* and *Desulfovibrio* sp. has been described by Cabrera et al. (2006). Removal of heavy metals in short-term bench scale upflow anaerobic packed bed reactor was described by Jong and Parry (2003), and the influence of copper and zinc on the mixed SRB population was studied by Utgikar et al. (2003). Rafida (2008) notes the significant role of biofilm formed during SRB activity.

3.4. CO-EXISTENCE AND COMPETITIONS BETWEEN OF THE DIFFERENT GROUPS OF MICROORGANISMS UNDER ANAEROBIC CONDITIONS

Anaerobic organic sewage treatment is based mainly on the activity of several groups of bacteria: fermentation, acetogens, methanogens, sulphate reducing bacteria and denitrifying bacteria. Some role is played also by bacteria reducing iron and oxygenated forms of other metals, but the content of these bacteria is very low.

The most well known and commonly applied method of anaerobic sewage treatment uses the activity of the first three mentioned microorganism groups. Each microorganism group partly oxygenates an organic compound to the relevant end products, which are next assimilated by the next link of the food chain till complete oxygenation (according to the scheme in Figure 4).

This system of biological sewage treatment in the biodegradation of organic pollutants begins with the activity of fermentation bacteria, which are responsible for hydrolysis and fermentation of particular organic compounds representing the main pollutant in the sewage. Many microorganisms display their ability to fermentation, in course of which various organic compounds are formed becoming an easily accessible carbon source for SRB.

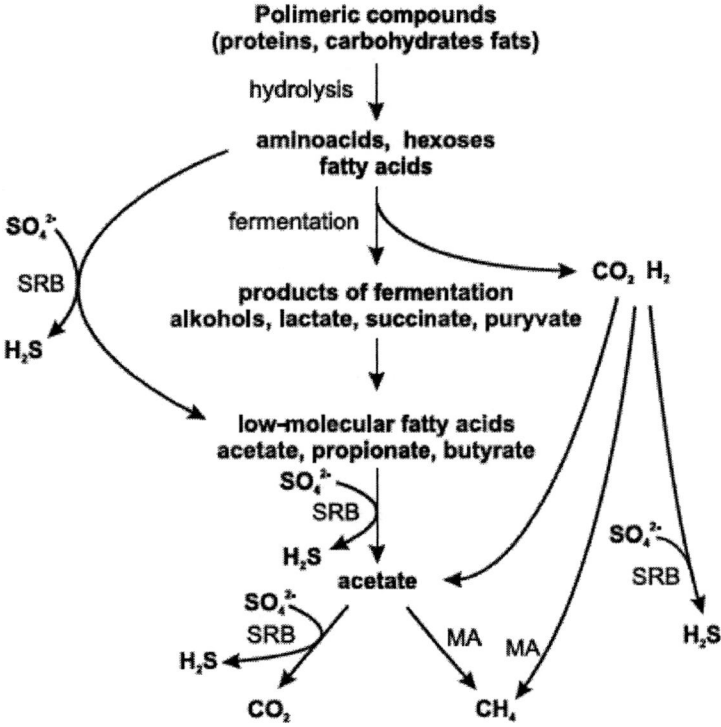

Figure 4. Possible stages of organic compounds biodegradation by SRB and methanogens. MA – methanogenic Archea, SRB – sulphate reducing bacteria.

The product of acetic acid fermentation, carried out by acetogens, is acetic acid. This reaction is catalyzed by enzymes produced mainly by acetate bacteria, but takes place also during other biochemical transformations.

$$CH_3CH_2OH + O_2 \rightarrow CH_3COOH + H_2O + 490 \text{ kJ } (118 \text{ kcal})$$

Various acetate bacteria species display smaller or larger abilities of further oxidation of acetic acid according to the following formula:

$$CH_3COOH + 2O_2 \rightarrow 2CO_2 + 2H_2O$$

The ability of lactic acid fermentation occurs in bacteria of the proper lactic acid fermentation. These bacteria oxidize simple carbohydrates and disaccharides to lactic acid (as the main product) and to various by-products. Lactic acid fermentation was observed e.g. in *Micrococcus*, *Escherichia* and *Microbacterium*. This fermentation can be sub-divided into two types: homofermentation – where lactic acid is the main product and there is a low content of by-products; and heterofermentation – where more by-products are formed in relation to lactic acid.

Lactic acid fermentation is a typical case of metabiosis i.e. growth of one group of microorganisms after another. In the first stage heterofermentation bacteria develop that acidify the environment and create favourable conditions for the development of bacteria from proper lactic acid fermentation (homofermentation). Ethanol is also one of the by-products of homofermentation, resulting from the anaerobic decomposition of carbohydrates into ethylic alcohol and carbon dioxide.

The products of butyric acid fermentation may be acetic acid, succinic acid and ethanol. This fermentation is carried out by *Clostridium* (*C. acetobutylicum*, *C. butylicum*), during which butanol is formed instead of butyric acid, and acetone instead of acetic acid. At first this fermentation is identical with butyric acid fermentation. However, when the produced acids decrease the reaction to about 4, fermentation changes course and neutral compounds such as butanol and acetone are formed instead of acids. There are also bacteria species which reduce acetone to isopropyl alcohol. It is worth noting that all fermentation products are an easily accessible carbon source for SRB.

The last stage of biological sewage treatment is the activity of methanogens. Using hydrogen, carbon dioxide and acetate produced by acetogens, they produce methane. It is estimated that about 1/3 methane produced is the product of CO_2 reduction, and the rest – of acetate decarboxylation.

The two main advantages of this treatment method is the formation of low quantities of excessive sludge, difficult to utilize, and the possibility of using the formed biogas as fuel. The energetic value of biogas increases with the content of methane, but it should be remembered that the content of methane in biogas is inversely proportional to the concentration of SO_4^{2-} in the treated sewage. Sulphates that are present in variable content in organic sewage do not directly influence the activity of methanogens, but they favour the selection of SRB, which according to reaction stoichiometry transform SO_4^{2-} to H_2S. H_2S has toxic influence on methanogens, and decreases the quality of biogas.

SRB have been noted in all types of bioreactors purifying sewage in anaerobic conditions. In such conditions they compete with various bacteria groups for the available organic compounds at almost all decomposition stages except hydrolysis, because most microorganisms capable of dissimilative sulphate reduction do not develop hydrolytic enzymes. The only exception is *Archeoglobus fulgidus*. In reactors treating high-sulphur sewage e.g. from pulp and paper, textile, pharmaceutical, metallurgic, paint and varnish, and plastics industries SRB are solely responsible for the final stages of organic pollutants decomposition (Colleran et al., 1995).

Some SRB species such as *Desulfuromonas acetoxidans* may form syntrophic relationships with photosynthesizing bacteria belonging to green sulphur bacteria (*Chlorobiaceae*). Under sun light the phototrophs assimilate carbon dioxide and oxidize hydrogen sulphide, which is the electron donor, to elemental sulphur that is released from the cell. In the following stage *Desulfuromonas* reduces it to hydrogen sulphide, with simultaneous oxidation of acetate and production of carbon dioxide. In some cases syntrophic relationships have been noted in other SRB species, e.g. *Desulfovibrio*. The common development of these bacteria is an example of a syntrophic relationship, in which substrates are distributed in two directions The product of the activity of one bacteria group becomes the substrate for the next group (Figure 5).

As mentioned above, SRB co-occur in all types of bioreactors applied in anaerobic treatment of sewage along with methanogens. The growth and increase of SRB activity depends on winning the rivalry with other bacteria groups on each stage of biodegradation. Competition of various microorganism groups in the presence of sulphates may take place in several different stages of the biodegradation process (Colleran et al., 1995):

- in the first stage of biodegradation competition with fermentation bacteria for products of biodegradation of polymeric compounds i.e. for simple monomeric compounds;

- competition with acetogens producing hydrogen for indirect fermentation products such as propionate or ethanol, which are an optimal carbon source for most SRB;
- competition with homoacetogens producing acetate for hydrogen, and at the end of biodegradation with methanogens for hydrogen or acetate.

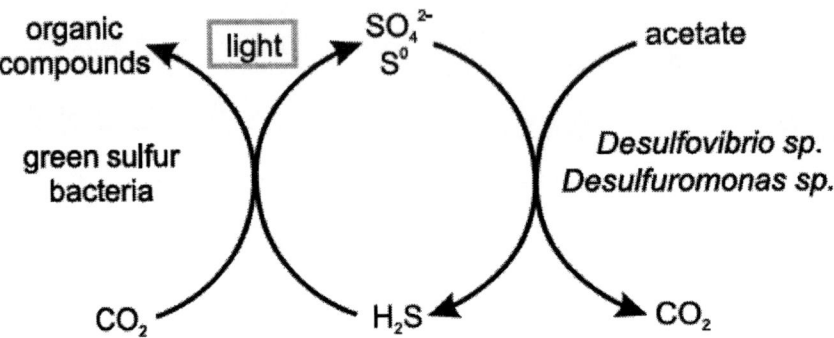

Figure 5. Sulphur cycle in a syntrophic relationship of a *Desulforomonas/Desulfovibrio* culture with green sulphur bacteria.

During sewage treatment conditions can be created to favour the development of a specific microorganism group. The dominating group of microorganisms depends on the concentration of sulphates in sewage and the chemical oxygen demand (COD). The calculated COD/SO_4 ratio in sewage supplies information about the possible selection of a specific group of microorganisms (Hao et al., 1996). If the value equals 0.67 or is lower, SRB dominate in the system and the decomposition end-products is not a mixture of CO_2 and CH_4, but CO_2 and H_2S. It seems that the co-existence of the two groups of microorganisms is possible in the range of COD/SO_4 from 1.7 to 2.7; below 1.7 SRB prevail (Clancy et al., 1992). If the ratio is higher, then more organic compounds are biodegraded during methanogenesis. It is estimated that at $COD/SO_4 > 10$ sulphides are not produced (Oude Elferinck et al., 1998). With known concentration of both components, i.e. TOC and $S-SO_4$ precise conditions for the development of a particular microorganism group can be defined and repeated (Oude Elferinck et al., 1998).

Decomposition of organic matter with the use of SRB is becoming to be applied in the treatment of high-sulphur waste. In comparison with the method based on methanogenesis, its advantage is the shorter duration of biodegradation, precipitation of heavy metals into poorly soluble sulphides and elimination of pathogenic organisms (toxic activity of hydrogen sulphide). The need to oxygenate hydrogen sulphide to elemental sulphur and slightly higher volume of excessive sludge than in methanogenesis is on the other hand a disadvantage.

3.5. BIOLOGICAL METHODS OF SULPHIDE TREATMENT – BY-PRODUCTS OF SULPHATE REDUCTION (PHOTOSYNTHESIZING AND CHEMOAUTOTROPHIC BACTERIA)

The process of anaerobic sewage treatment using sulphidogenesis is linked with the necessity to remove hydrogen sulphide produced by SRB. This requirement seems the most important disadvantage of the method; therefore many reports regarding this topic have been published (Kobayashi et al., 1983; Khanna et al., 1996; Lee and Kim, 1998; Tichy et al., 1998). So far, the often proposed solution is chemical (abiotic) oxidation of hydrogen sulphide to elemental sulphur; this process is, however, very expensive due to the application of catalysts and energy demanding aeration. Therefore, solutions with application of microorganisms capable of replacing chemical removal of hydrogen sulphide are sought for. It seems that the most appropriate microorganisms are photosynthesizing and chemolithoautotrophic bacteria. Both groups comprise microorganisms capable to oxidation of hydrogen sulphide, although the metabolism linked with this process is entirely different.

The most important problem during application of microorganisms in removing hydrogen sulphide is the selection of tribes and physical-chemical conditions producing elemental sulphur in course of oxidation; the sulphur would then be emitted from the reactor. It is thus indispensable that the final effect of sulphur transformation in anaerobic high-sulphur sewage treatment using sulphidogenesis and additional refinery stages would be elemental sulphur being a potential source in many industrial branches. It poses much less hazard to the natural environment during stacking than other solid waste containing sulphur, e.g. waste gypsum.

3.5.1. Microorganisms Applied in Biological Hydrogen Sulphide Treatment

Photosynthesizing sulphur bacteria. This is the largest and probably the most appropriate group with regard to application in hydrogen sulphide removal that encompasses microorganisms capable of oxidizing hydrogen sulphide to elemental sulphur (Gemerden, 1986; Eraso and Kaplan, 2001). It comprises a diverse group of microorganisms classified into many taxonomic units which share the ability to carry out processes depending on presence of light. Typically, four sub-groups are distinguished: green sulphur bacteria, purple sulphur bacteria, purple non-sulphur bacteria and the *Chloroflexus* group comprising non-sulphur green bacteria. Due to their ability to cumulate elemental sulphur the most appropriate for application in biotechnology of hydrogen sulphide oxidation are the first two groups.

Green sulphur bacteria conduct photosynthesizing processes in which the electron sources include reduced sulphur compounds such as sulphides and tiosulphate. These bacteria contain high amounts of bacteriochlorophyll c, d, e, as well as much lower quantities of bacteriochlorophyll a (Brock and Madigan, 2006; Baneras et al., 1999). They are assigned to the order *Chlorobiales*, which comprises such species as *Chlorobium limicola*, *Chlorobium limicola thiosulfatophilum*, and *Pelodictyon* sp. They occur in aqueous environs, particularly in thermally stratified lakes, but can also be found in saline conditions or in highly exceeded temperatures. Many species of green bacteria are easily isolated from the natural environment and multiplied in laboratory conditions, what allows potential application in biotechnology. The most important feature of the group with regard to this fact is the ability to cumulate elemental sulphur outside the cells, what distinguishes them from another group of photosynthesising bacteria – purple sulphur bacteria of potential significance in the process.

Microorganisms belonging to purple sulphur bacteria contain high amounts of bacteriochlorophyll a and have the ability to cumulate sulphur, but only within the cell, with the exception of the family *Ectotiorhodospiraceae* that cumulates sulphur outsides the cell (Eraso and Kaplan, 2001). The basic source of electrons in the photosynthesizing processes includes hydrogen and hydrogen sulphide, and the stored sulphur may potentially be oxidized to sulphate. The bacteria belong to the order *Chromatiales*, and the best known species is *Chromatium okeni*.

Chemolithoautotrophic sulphur bacteria. The second large group potentially applicable in biotechnology of hydrogen sulphide removal comprises autotrophic bacteria carrying out oxidation processes of reduced inorganic compounds such as sulphides, hydrogen sulphide, iron (II) as well as tiosulphate and elemental sulphur. The processes supply to the bacteria cells energy needed to bind carbon dioxide. The group includes such species as *Acidithiobacillus tiooxidans, A. ferrooxidans, Thiobacillus thioparus,* and *Starkeya novella*. Chemolithoautotrophic sulphur bacteria are generally applied in processes of bioleaching of metals from low-percentage ores or other waste containing metal sulphides, as well as in bioremediation (Third et al., 2000; Suzuki, 2001; Pacholewska, 2004; Suzuki and Suko, 2006). Due to this fact these bacteria could theoretically be appropriate to apply in recovery of metals from the remaining deposit, e.g. after utilization of phosphogypsum during sulphidogenesis. Such deposit, due to heavy metal content in phosphogypsum would have a high amount of insoluble metal sulphides that potentially could be utilized. However, application of chemolithoautotrophic bacteria in refinement of sewage rich in hydrogen sulphide has created a number of problems. First of all, despite the existence of species producing sulphur such as *Thiobacillus thioparus*, a large number of bacteria species oxidize hydrogen sulphide directly to sulphates. Secondly, these bacteria often require lower pH in the environment, and thirdly, the effluent after refinement using sulphidogenesis may contain organic compounds distinctly hampering the activity of chemolithoautotrophic sulphur bacteria. Due to this fact, the present application of these bacteria is restricted to attempts of metal bioleaching and in refinement of mineral sewage containing e.g. tiosulphates.

3.5.2. Application of Photosynthesizing Sulphur Bacteria in Removal of Hydrogen Sulphide from Sewage

Attempts to apply green and purple sulphur bacteria in the removal of hydrogen sulphide are justified in all processes that generate the formation of sewage strongly contaminated by sulphides, hydrogen sulphide and tiosulphates. Such sewage is formed during gas desulphurization, in crude oil processing plants, chemical plants, and mostly during treatment of high-sulphur sewage using aerobic methods. Solutions proposed hitherto are based on the utilization of photosynthesizing bacteria populations in photoreactors, in which are ensured conditions favourable for their growth. The most important and practically the only necessity is the assurance of a light source, what is linked with some cost, but due to the possibility of using day-light, the cost is distinctly lower than in the case of a chemical process. Lee and Kim (1998) proposed the application of the so-called optical-fibre bioreactor, to which light is brought by optical fibres. The bacteria *Chlorobium limicola thiosulfatophilum* were used in the bioreactor with successful results. Henshaw et al. (1998) also applied *Chlorobium limicola* in a system similar to a chemostat, in a suspended-growth continuous stirred tank reactor, which was illuminated by a lamp emitting infrared light. The process was highly effective, with 90% conversion of sulphides in the solution to elemental sulphur. Henshaw et al. (1999) tested the influence of the material used to construct the bioreactor (mainly various types of plastics permitting infrared radiation) and showed the lack of significant influence of the applied material on the growth of *Chlorobium limicola*. It is worth noting that a rather low light intensity was applied in this experiment, from $3.4 \cdot 10^{-3}$ to $4.7 \cdot 10^{-3}$ W m^{-2}.

Application of photosynthesizing bacteria is linked with the important issue of organic compounds presence in sewage flowing to the reactor. Due to anaerobic conditions SRB may grow in the reactor; at this stage of refinement this is highly undesirable, influencing the effective reduction of hydrogen sulphide pollution. Theoretically, the presence of organic compounds may be unfavourable to the metabolism of photosynthesizing bacteria, although it seems that they are partly capable to utilize simple organic compounds. This problem is crucial during attempts of applying photosynthesizing bacteria to remove hydrogen sulphide from sewage earlier treated during sulphidogenesis. Such sewage may contain also indistinct amounts of organic matter and may be also the source of SRB. Borkowski and Wolicka (2007a) applied an assemblage of photosynthesizing bacteria isolated from the natural environment on a flooded deposit and indicated that yeast extract flowing with synthetic medium may distinctly inhibit the efficiency of sulphide oxidation. In turn, in Borkowski and Wolicka (2007b) the synthetic medium flowing onto the flooded deposit colonized by photosynthesizing bacteria was replaced by filtered influent from a sulphidogenic bioreactor, in which phosphogypsum with distillery decoctions was treated. In this case, 60% effectiveness of sulphide content reduction was obtained (from 163 mg L^{-1} to 70 mg L^{-1}) with partial conversion to elemental sulphur.

During simultaneous treatment of organic sewage and solid gypsum waste such as phosphogypsum using anaerobic treatment by sulphidogenesis, it is possible to refine the sewage already after the main process (Figure 6).

The process should comprise the following stages:

1. The effluent from the anaerobic reactor after sulphidogenesis contains sulphides and hydrogen sulphide; after primary filtering it is transported to a bioreactor with photosynthesizing bacteria. The effluent can contain also organic compounds and SRB, the number of which should distinctly decrease after filtration. The effect is almost complete reduction of the remaining organic pollutants and elemental sulphur.
2. The solid waste after phosphogypsum may contain high amounts of calcite and slightly less phosphates and due to this fact may be applied in agrotechnical activities. However, if the phosphogypsum contained a high content of heavy metals or if these metals flew with sewage, after

utilization using sulphidogenesis the deposit is enriched in sulphides of heavy metals. Such deposit, beside the fact that it is very toxic, is also a significant source of metal that can be recovered during bioleaching by natural or selected communities of chemolithoautotrophic sulphur bacteria representing e.g. *Thiobacillus* and *Acidithiobacillus*.

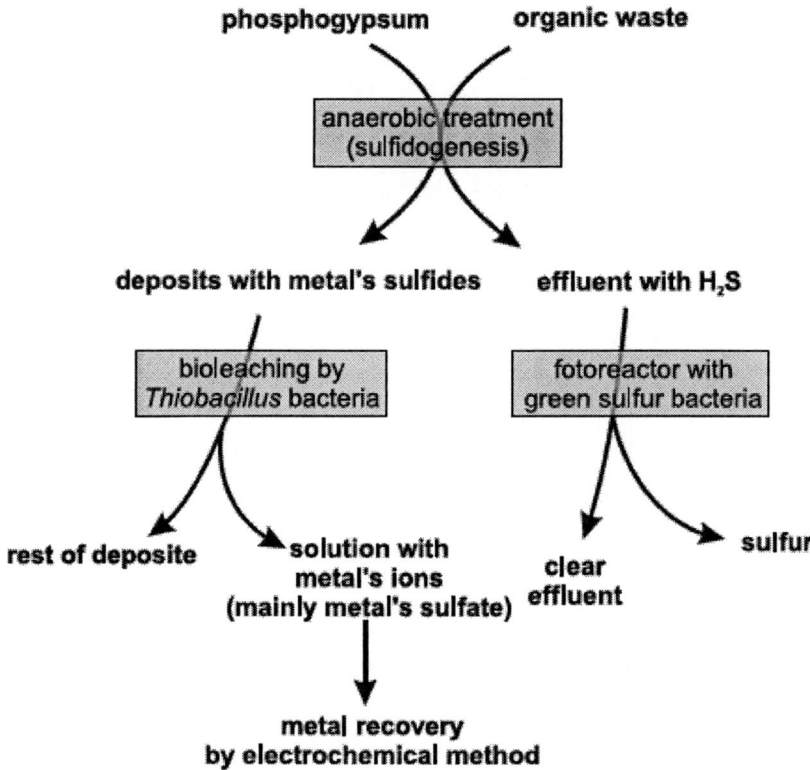

Figure 6. Biological oxidation of hydrogen sulphide after sulphidogenesis.

Conclusion

At present it is obvious that natural environment protection is a basic and essential task. Taking into account the great volume of sewage produced in anthropogenic processes by industrial plants, it is important to find solutions aimed at decreasing the influence of toxic and dangerous sewage on the natural environment. One of the basic methods of solving the water quality issue is rational and professional approach to problems linked with the treatment of various types of organic sewage.

It is commonly known that biological methods using microorganisms can be applied in the biodegradation of hazardous petrochemical waste. Many groups of anaerobic bacteria are able to biodegrade various types of organic waste to non-toxic compounds or even to inorganic compounds after complete mineralization. Biological methods do not require introduction of chemical compounds to the environment that would negative influence the biocenosis. Most of all, biological methods take place naturally in the environment, and anthropogenic influence in this case is focused only on stimulating their range and intensity. Final products of microbiological decomposition include carbon dioxide and water.

Simultaneous biodegradation of several industrial wastes, e.g. organic sewage and solid waste seems an interesting and optimal solution from the point of economy. Costs linked with simultaneous biodegradation of two arduous industrial wastes are almost always much lower than separate treatment processes.

Application of SRB in biological treatment of sewage is becoming popular and is entirely justified.

ACKNOWLEDGMENTS

I would like to thanks Dr Andrzej Borkowski for his help in preparing of this manuscript.

REFERENCES

Aeckersberg, F. Rainey, F. A., Widdel, F. (1998). Growth, natural relationship, cellular fatty acids and metabolic adaptation of sulfate-reducing bacteria that utilize long-chain alkanes under anoxic conditions. *Archives of Microbiology, 170*, 361-369.

Baena, S. Fardeau, M. L. Labat, M. Ollivier, B. Garcia, J. L., Patel, B. K. C. (1998). *Desulfovibrio aminophilus* sp.nov., a novel amino acid degrading and sulfate reducing bacterium from an anaerobic dairy wastewater lagoon. *Systematic and Applied of Microbiology, 21*, 498-504.

Baena, S. Fardeau, M. L. Labat, M. Ollivier, B. Garcia, J. L., Patel, B. K. (2000). *Aminobacterium mobile* sp. nov., a new anaerobic amino-acid-degrading bacterium. *International Journal of Systematic Bacteriology, 1*, 259-264.

Baena, S. Fardeau, M. L. Ollivier, B.M. Labat, M. Thomas, P. Garcia, J. L. Patel, B. K. C. (1999). *Aminomonas paucivorans* gen. nov., sp. nov., a mesophilic, anaerobic, amino-acid-utilizing bacterium. *Journal Int J Syst Bacteriol, 49*, 975-982.

Bak, F. and Widdel, F. (1986). Anaerobic degradation of phenol and phenol derivatives by *Desulfobacterium phenolicum*. *Archives of Microbiology, 54*, 177-180.

Ball, H. A. and Reinhard, M. (1996). Monoaromatic hydrocarbon transformation under anaerobic conditions at Seal Beach, California: labolatory studies. Environmental *Toxicology and Chemistry, 15*, 114-122.

Bañeras, L. Borrego, C. M. Garcia-Gil, L. J. (1999). Growth-rate-dependent bactriochlorophyll c/d ratio in the antenna of *Chlorobium limicola* strain UdG6040. *Arch Microbiol, 171*, 350-354.

Beller, H. R. Garbić- Galić, D. Reinhard, M. (1992). Microbial degradation of toluene under sulfate-reducing conditions and the influence of ion on the process. *Applied and Environmental Microbiology, 58*, 786-793.

Boopathy, R. Gurgas, M. Ullian, J. Manning, J. F. (1998). Metabolisms of explosive compounds by sulfate-reducing bacteria. *Current Microbiology, 37*, 127-131.

Borkowski, A. and Wolicka, D. (2007a). Geomicrobiological aspects of the oxidation of reduced sulfur compounds by photosynthesizing bacteria. *Polish Journal of Microbiology, 56(1)*, 53-57.

Borkowski, A. and Wolicka, D. (2007b). Isolation and characteristics of photosynthesizing bacteria and their utilisation in sewage treatment. *Polish Journal of Environmental Studies, 16(3B)*, 38-42.

Bothe, H. and Trebst, A. (1981). Nitrogen; sulphur; microbial metabolism; plants; sulfur; nitrogen fixation; metabolism; Fixation. Springer Verlag, Berlin and New York.

Brock, T. D. and Madigan, M. T. (2006). Biology of microorganisms. Prentice-Hall International, Inc. *Chapter 16*, 562-608.

Cabrera, G. Pérez, R. Gómez, J. M. Ábalos, A. Cantero, D. (2006). Toxic effects of dissolved heavy metals on *Desulfovibrio vulgaris* and *Desulfovibrio sp.* strains. *Journal of Hazardous Metals*, 40-46.

Caldwell, M. E. Tanner, R. S. Suflita, J. M. (1999). Microbial metabolisms of benzene and the oxidation of ferrous iron under anaerobic conditions: implications for bioremediation. *Anaerobe, 5*, 595-603.

Chen, C. I. and Taylor, R. T. (1997). Thermophiloc biodegradation of BTEX by two consortia of anaerobic bacteria. *Applied Microbiology and Biotechnology, 48*, 121-128.

Chi Ming, S. and Young, L. Y. (1999). Isolation and characterization of a sulfate-reducing bacterium that anaerobically degrades alkanes. *Applied and Environmental Microbiology, 65*, 2969-2976.

Clancy, P. C. Venkataraman, N. Lynd, L. R. (1992). Biochemical inhibition of sulphate reduction in batch and continuous anaerobic digesrters. *Water Science and Technology, 25*, 51.

Colleran, E. Finnegan, S. Lens, P. (1995). Anaerobic treatment of sulphate – containing waste streams. *Antonie van Leeuwenhoeck, 67*, 29-46.

Deswaef, S. Salmon, T. Hiligsmann, S. Taillieu, X. Milande, N. Thonart, P. Cruine, M. (1996). Treatment of gypsum waste in a two stage anaerobic reactor. *Water Science and Technology, 34*, 367-374.

Dilling, W. and Cypionka, H. (1990). Aerobic respiration in sulfate-reducing bacteria. *FEMS Microbiology Letters, 71*, 123-128.

Dvorak, D. H. Hedin, R. S. Edenborn, H. M. McIntire, P. E. (1992). Treatment of metal contaminated water using bacterial sulfate reduction: results from pilot-scale reactors. *Biotechnology and Bioengineering, 40*, 609-616.

Edwards, E .A. Wills, L. E. Reinhard, W. Garbić-Galić, D. (1992). Anaerobic degradation of toluene and xylene by aquifer microorganisms under sulfate-reducing conditions. *Applied and Environmental Microbiology, 58*, 794-503.

Edwards, E. A. and Garbić-Galić, D. (1992). Complete mineralization of benzene by aquifer microorganisms under strictly anaerobic conditions. *Applied and Environmental Microbiology, 58*, 2663-2666.

Ekstrom, E. B. and Morel, F. M. M. (2008). Cobalt limitation of growth and mercury methylation in sulfate-reducing bacteria. *Environmental Science and Technology, 42*, 93-99.

Eraso, J. M. and Kaplan, S. (2001) Photoautotrophy. Encyclopedia of Life Sciences, Nature Publishing Group.

Fauque, G. Legall, J. Barton, L. L. (1991). Sulfate-reducing and sulfur reducing bacteria. Variations in autotrophic life. In J. M. I Shively, and L. L. Barton (Eds.), Academic Press Ltd.

Galusho, A. S. and Rozanova, E. P. (1991). *Desulfobacterium cetonicum* sp. nov.: a sulfate-reducing bacterium which oxidizes fatty acids and ketones. *Mikrobiologia, 60*, 742-746.

Gemerden, H. (1986). Production of elemental sulfur by green and purple sulfur bacteria. *Arch Microbiol, 146*, 52 – 56.

Gibert, O. de Pablo, J. Cortina, J. L. Ayora, C. (2002). Treatment of acid mine drainage by sulphate-reducing bacteria using permeable reactive barriers: A review from laboratory to full-scale experiments. *Reviews in environmental Science and Biotechnology, 1*, 327-333.

Gibson, G. (1990). Physiology and ecology of the sulphate-reducing bacteria. *Journal of Applied Bacteriology, 69*, 769-797.

Gieg, L. M. and Suflita, J. M. (2002). Detection of anaerobic metabolites of saturated and aromatic hydrocarbons in petroleum-contaminated aquifers. *Environmental Science and Technology, 36*, 3755-3762.

Hao, O. J. Chen, J. M. Huang, L. Buglass, R. L. (1996). Sulfate-reducing bacteria. *Critical Reviews in Environmental Science and Technology, 26*, 155-187.

Harms, G. Zengler, K. Rabus, R. Aeckersberg, F. Minz, D. Rossello-Mora, R. Widdel, F. (1999). Anaerobic oxidation of o-xylene, and homologous alkylbenzenes by new types of sulfate-reducing bacteria. *Applied and Environmental Microbiology, 65*, 999-1004.

Hass, C. and Polpraset, C. (1993). Biological sulfide prestripping for metal and COD removal. *Water Environment Research, 65*, 645-649.

Henshaw, P. F. Bewtra, J. K. Biswas, N. (1998). Hydrogen sulphide conversion to elemantal sulphur in a suspended-growth continuous stirred tank reactor using *Chlorobium limicola. Water Research, 32(6),* 1769-1778.

Henshaw, P. Medlar, D. Mcewen, J. (1999). Selection of a support medium for a fixed-film green sulphur bacteria reactor. *Water Research, 33(14)*, 3107-3110.
Hernandez-Eugenio, G. Fardeau, M. L. Patel, B. K. C. Garcia, J. L. Ollivier, B. (2000). *Desulfovibrio mexicanus* sp.nov., a sulfate-reducing bacterium isolated from an upflow anaerobic sludge blancet (UASB) reactor treating cheese wastewaters. *Anaerobe, 6*, 305-312.
Jenneman, G. E. and Gevertz, D. (1999). Identification, characterization and application of sulfide-oxidizing bacteria in oil fields. In Microbial Biosystems: New Frontiers. Proceedings of the 8 th International Symposium on Microbial Ecology. C. R. Bell, M. Brylinsky, and P. Johnson-Green (Eds.), Atlantic Canada Society for Microbial Ecology, Halifax, Canada.
Jong, T. and Parry, D. L. (2003). Removal of sulfate and heavy metals by sulfae reducing bacteria in short-term bench scale upflow anaerobic packed bed reactor runs. *Water Research, 37*, 3379-3389.
Karnachuk, O. V. Kurochkina, S. Y. Nicomrat, D. Frank, Y. A. Ivasenko, D. A. Phyllipenko, E. A. Tuovinen, O. H. (2003). Copper resistance in *Desulfovibrio* strain R2. *Antonie van Leeuwenhoek International Journal of General and Molecular Microbiology, 83*, 99–106.
Karnachuk, O. V. Kurochkina, S. Y. Tuovinen, O. H. (2002). Growth of sulfate-reducing bacteria with solid-phase electron acceptors. *Applied Microbiology and Biotechnology, 58*, 482-486.
Kaufman, E. N. Little, M. H. Selvaraj, P. T. (1996). Recycling FGD gypsum to calcium carbonate and elemental sulphur using mixed sulfate reducing bacteria with sewage digest as a carbon source. *Journal of Chemical Technology and Biotechnology, 66*, 365.
Khanna, P. Rajkumar, B. Jothikumar, N. (1996). Microbial recovery of sulfur from thiosulfate-bearing wastewater with phototrophic and sulfur-reducing bacteria. *Current Microbiol., 32*, 33-37.
Kleikemper, J. Pelz, O. Schroth, M. H. Zeyer, J. (2002). Sulfate-reducing bacterial comminity response to carbon source amendments in contaminated aquifer microcosms. *FEMS Microbiology Ecology, 42*, 109-118.
Kniemeyer, O. Fisher, T. Wilknes, H. Glőckner, Widdel, F. (2003). Anaerobic degradation of ethylbenzene by a new type of marine sulfate-reducing. *Applied and Environmental Microbiology, 69*, 760-768.
Kobayashi, H. A. Stenstrom, M. Mah, R. A. (1983). Use of photosynthetic bacteria for hydrogen sulfide removal from anaerobic waste treatment effluent. *Water Research, 17(5)*, 579-587.

Koschorreck, M. (2008). Microbial sulphate reduction at a low pH. *FEMS Microbiology Ecology, 64*, 329-342.

Kowalski, W. (2002). The kinetic of microbiological and mineralogical processes in phosphogypsum utilisation. Project report KBN 9T12B00515 (in Polish).

Kowalski, W. Przytocka-Jusiak, M. Błaszczyk, M. Hołub, W. Wolicka, D. Wesołowska, I. (2002). Effect of nitrate on biotransformation of phosphogypsum and phenol uptake in cultures of autochtonous sludge mikroflora from petroleum refining wastewaters. *Acta Microbiologica Polonica, 51*, 47.

Kuever, J. Rainey, F. A. Hippe, H. (1999). Description of *Desulfotomaculum sp.*Groll as *Desulfotomaculum gibsoniae* sp.nov. *International Journal of Systematic Bacteriology, 49*, 1801-1808.

Kutera, J. and Talik, B. (1996). Oczyszczanie ścieków z mleczarni ze szczególnym uwzględnieniem metod naturalnych w środowisku glebowo-roślinnym, *Zeszyty Nauk. AR Wrocław*, Konf. XIII, Nr 293, 1996.

Lee, K. Kim, and B.W. (1998). Enhanced microbial removal of H_2S using *Chlorobium* in a optical-fiber bioreactor. *Biotechnology Letters, 20(5)*, 525-529.

Lens, P. N. L. van den Bosch, M. C. Hulshoff Pol, L. W. Lettinga, G. (1998). Effect of staging on volatile fatty acid degradation in a sulfidogenic granular sludge reactor. *Water Research, 32(4)*, 1178-1192.

Lesage, S. Li, W. C. Millar, K. Brown, S. Liu, D. (2000). Effect of humic acids on of bioremediation of polycyclic aromatic hydrocarbons from aquifers contaminated with petroleum. http://biogroup.gzea.com/bioreferences/tier 1papers/effect.asp.

Lovley, D. R. Coates, J. D. Woodward, J. C. Phillips, E. J. P. (1995). Benzene oxidation coupled to sulfate reduction. *Applied and Environmental Microbiology, 61*, 953-958.

Magot, M. Ollivier, B. Patel, B. K. C. (2000). Microbiology of petroleum reservoirs. *Antonie van Leeuwenhoek International Journal of General and Molecular Microbiology, 77*, 103.

Mizuno, O. Takagi, H. Noike, T. (1998). Biological sulfate removal in an acidogenic bioreactor with an ultrafiltration membrane system. *Water Science and Technology, 38*, 4-5.

Mueller, F. R. and Nielsen, P. H. (1996). Characterization of thermophilic consortia from two souring oil reservoirs. *Applied and Environmental Microbiology, 62*, 3083.

Mueller, R. F. and Steiner, A. (1992). Inhibition of anaerobic digestion caused by heavy metals. *Water Science and Technology, 26*, 835-846.

Oude Elferinck, S. J. W. H. Vorstman, W. J. C. Sopjes, A. Stams, A. J. M. (1998). *Desulforhabdus amnigenus* gen.sp.nov., a sulfate reducers isolated from anaerobic granular sludge. *Archives of Microbiology, 164,* 119-124.

Pacholewska, M. (2004). Bioleaching of galena flotation concentrate. *Physicochemical Problems of Mineral Processing, 38,* 281-290.

Postgate, J. R. (1984). The sulphate reducing bacteria. Cambridge University Press.

Przytocka-Jusiak, M. Rzeczycka, M. Ponichtera, E. Mycielski, R. (1997). Rozkład benzenu w hodowlach termofilnych bakterii redukujących siarczany. In I cześć materiałów V Ogólnopolskiego Sympozjum Naukowo-Technicznego „Biotechnologia Środowiskowa" Ustroń-Jaszowiec, pp. 215-222.

Rabus, R. Nordhaus, R. Ludwig, W. Widdel, F. (1993). Complete oxidation of toluene under strictly anoxic conditions by a new sulfate-reducig bacterium. *Applied and Environmental Microbiology, 59,* 1444-1451.

Rafida, A. I. (2008). Removal of heay metals from rainwater in batch reactors with sulphate reducing bacteria (SRB). *Proceeding of World Academy of Science, Enegineering and Technology, 33,* 23-27.

Rees, G. N. Harfoot, C. G. Sheehy, A. J. (1997). Amino acid degradation by the mesophilic sulfate-reducing bacterium *Desulfobacterium vacuolatum*. *Archives of Microbiology, 169,* 76-80.

Reichenbecher, W. and Schink, B. (1997). *Desulfovibrio inopinatus*, sp. nov., a new sulfate-reducing bacterium that degrades hydroxyhydroquinone (1,2,4-trihydroxybenzene). *Archives of Microbiology, 168,* 338-344.

Rozanova, E. P. Borzenkov, I. A. Tarasov, A. L. Suntsova, L. A. Dong, Ch. L. Belyaev, S. S. Ivanov, M. V. (2001). Microbiological processes in a high-temperature oil field. *Microbiology, 70,* 102-110.

Rueter, P. Rabus, R. Wilknes, H. Aeckersberg, F. Rainey, F., A. Jannasch, H. W. Widddel, F. (1994). Anaerobic oxidation of hydrocarbons in crude oil by new types of sulphate-reducing bacteria. *Nature, 372,* 455-457.

So, C. M. and Young, L. Y. (1999). Initial reactions for anaerobic alkane degradation by the sulfate reducer strain AK-01. *Applied and Environmental Microbiology, 65,* 5532-5540.

Suzuki, I. (2001). Microbial leaching of metals from sulfide minerals. *Biotechnology Advances, 19,* 119-132.

Suzuki, Y. and Suko, T. (2006). Geomicrobiological factors that control uranium mobility in the environment: update on recent advances in the bioremediation of uranium-contaminated sites. *Journal of Mineralogical and Petrological Sciences, 101,* 299-307.

Szewzyk, R. and Pfenning, N. (1987). Complete oxidation of catechol by stricly anaerobic sulfate reducing *Desulfobacterium catecholicum* sp.nov. *Archives of Microbiology, 147*, 163-168.

Third, K. A. Cord-Ruwisch, R. Watling, H. R. (2000). The role of iron-oxidizing bacteria in stimulation or inhibition of chalcopyrite bioleaching. *Hydrometallurgy, 57*, 225-233.

Tichy, R. Janssen, A. Grotenhuis, J. T. C. van Abswoude, R. Lettinga, G. (1998). Oxidation of biologically-produced sulphur in a continuous mixed-suspension reactor. *Water Research, 32(3)*, 701-710.

Utgikar, V. P. Tabak, H. H. Haines, J. R. Govind, R. (2003). Quantification of toxic and inhibitory impact of copper and zinc on mixed cultures of sulfate-reducing bacteria. *Biotechnology and Bioengineering, 82*, 306-312.

Voordouw, G. Armstrong, S. M. Reimer, M. F. Fouts, B. Telang, A. J. Shen, Y. Gevertz, D. (1996). Characterization of 16S rRNA genes from oil field microbial communities indicates the presence of a variety of sulfate-reducing, fermentative, and sulfide-oxidizing bacteria. *Applied and Environmental Microbiology, 62*, 1623-1629.

Warren, E. Bekins, B. A. Godsy, M. (1999). Inhibition of acetoclastic methanogenesis by crude oil from Bemidji, Minnesota.

Widdel, F. and Bak, F. (1992). Gram-negative mesophilic sulfate-reducing bacteria. In A. Balows, H. G. Truper, M. Dworkin, W. Harder, K. H. Schleifer (Eds.), *The prokaryotes*, 2nd end, chapter 12.

Wilkes, H. Boreham, C. Harms, G. Zengler, K. Rabus, R. (1999). Anaerobic degradation and carbon isotopic fractionation of alkylobenzenes in crude oil by sulphate-reducing bacteria. *Organic Geochemistry, 31*, 101-115.

Wit, R. (1992). Sulfide containing environments. In J. Ledeberg (Eds.), *Encyclopedia of Microbiology*, Academic Press, New York, 105-121.

Wolicka, D. (2006). Biotransformation of phosphogypsum in cultures of bacteria isolated from petroleum. In Proceedings of the International Conference Protection and Restoration of the Environmental VIII, Create, 2006. Nikas Books, 217-218.

Wolicka, D. (2008a). Biodegradation-the natura metod for liquidation enironmental polluted by petroleum products. *Oil and Gas Institute, Prace, 150*, 675-680.

Wolicka, D. (2008b). Biotransformation of phosphogypsum in wastewaters from dairy industry. *Bioresource Technology, 99*, 5666-5672.

Wolicka, D. and Borkowski, A. (2007a). Participation of sulfate-reducing bacteria in biodegradation of organic matter in soils contaminated with petroleum products. *Archives of Environmental Protection, 33*, 93-99.

Wolicka, D. and Borkowski, A. (2007b). Activity of a sulphate reducing bacteria community isolated from an acidic lake. EANA 07, 7th European Workshop on Astrobiology, Turku, Finland, October 22-24, 2007, pp.98.

Wolicka, D. and Borkowski, A. (2008). Oil-derived products as the source of sulphate-reducing bacteria biotransforming phosphogypsum. *Polish Journal of Environmental Studies, 17*, 592-595.

Wolicka, D. and Borkowski, A. (2009). Phosphogypsum biotransformation in cultures of sulphate reducing bacteria in whey. *International Biodeterioration and Biodegradation, 63*, 322-327.

Wolicka, D. and Kowalski, W. (2005). Utilization of different carbon compounds by sulphate-reducing bacteria in medium with phosphogypsum. *Archives of Environmental Protection, 31*, 105-112.

Wolicka, D. and Kowalski, W. (2006a). Biotransformation of phosphogypsum on distillery decoctions (preliminary results). *Polish Journal of Microbiology, 55*, 147-151.

Wolicka, D. and Kowalski, W. (2006b). Biotransformation of phosphogypsum in petroleum-refining wastewaters. *Polish Journal of Environmental Studies, 15*, 355-360.

Wolicka, D. Kowalski, W. Boszczyk-Maleszak, H. (2005). Biotransformation of phosphogypsum by bacteria isolated from petroleum-refining wastewaters. *Polish Journal of Microbiology, 2*, 54.

Zhang, W. and Bouwer, E. J. (19970. Biodegradation of benzene, toluene and naphtalene in soil–water slurry microcosms. *Biodegradation, 8*, 167-175.

INDEX

A

abiotic, 33
acceptor, 7, 11, 12, 13, 14, 17
acceptors, 8, 10, 12, 13, 46
acetate, 9, 10, 11, 12, 13, 23, 30, 31, 32
acetic acid, 8, 30
acetone, 20, 30
acid, 8, 10, 13, 21, 22, 30, 43, 45, 47, 48
acidic, 2, 11, 18, 21, 50
acidification, 28
adaptation, 24, 43
adsorption, 3, 20
aerobic, 4, 5, 15, 25, 35
aerobic bacteria, 15
agar, 14, 15
agents, 18, 20
agricultural, xi, 1
agriculture, 4, 5, 25
alcohol, 21, 30
alcohols, 7, 8, 18, 20, 21
aldehydes, 20, 21
aliphatic compounds, 9
ALK, 9
alkali, 20, 22
alkaline, 3, 21
alkane, 48
alkanes, 9, 43, 44
amendments, 46
amino acid, 7, 8, 43
amino acids, 7, 8
ammonium, 2
anaerobes, 7, 8, 15
anaerobic, 4, 5, 7, 8, 9, 12, 13, 14, 15, 24, 27, 28, 29, 30, 31, 33, 36, 39, 43, 44, 45, 46, 47, 48, 49
anaerobic bacteria, 14, 15, 39, 44
anaerobic granular sludge, 48
anaerobic sludge, 46
animals, 19
anoxic, 43, 48
antagonists, 12
antenna, 43
anthropogenic, xi, 1, 3, 4, 7, 22, 26, 39
antimony, 25
application, 2, 3, 4, 5, 15, 17, 18, 19, 20, 22, 23, 27, 33, 34, 35, 46
aquifers, 45, 47
archaea, 11, 15
aromatic compounds, 9
aromatic hydrocarbons, 7, 18, 21, 45
arsenic, 25
Atlantic, 46
atmosphere, 26
atmospheric pressure, 20
ATP, 10, 12
autotrophic, 34, 45

Index

B

bacteria, 4, 5, 7, 8, 10, 12, 13, 14, 15, 20, 23, 24, 25, 28, 29, 30, 31, 32, 33, 34, 35, 36, 37, 39, 43, 44, 45, 46, 48, 49, 50
bacterial, 44, 46
bacterium, 43, 44, 45, 46, 48
ballast, 19
barriers, 45
batteries, 26
benzene, 9, 15, 44, 45, 50
beverages, 19
biodegradable, 2, 18
biodegradation, 3, 4, 5, 9, 18, 20, 22, 23, 28, 29, 31, 32, 39, 44, 49
biofilms, 23
biogas, 31
biomass, 3, 24, 27
biopolymers, 27
bioreactor, 5, 24, 35, 36, 47
bioreactors, 7, 23, 24, 31
bioremediation, 9, 28, 35, 44, 47, 48
biosorption, 27
biosphere, xi, 1
biotechnology, 3, 4, 34
biotransformation, 3, 4, 5, 22, 24, 47, 49, 50
boiling, 20
breeding, 18, 19
butadiene, 20
butyric, 30
by-products, 2, 30

C

cadmium, 25
calcium, 4, 23, 46
calcium carbonate, 46
carbohydrates, 7, 9, 30
carbon, 4, 8, 9, 10, 11, 12, 13, 18, 20, 29, 30, 31, 32, 34, 39, 46, 49, 50
carbon dioxide, 10, 30, 31, 34, 39
carbonates, 4, 22
carboxylates, 7
casein, 14, 18, 19, 20

catalysis, 28
catechol, 9, 10, 49
cell, 13, 27, 28, 31, 34
cell membranes, 27
ceramic, 24
CH_3COOH, 12, 30
CH_4, 32
cheese, 18, 19, 46
chemical composition, 3, 4
chemical reactions, 27
chromium, 28
circulation, 23
classical, 3
clay, 24
cleaning, 18, 20
Co, 28
CO_2, xi, 1, 8, 10, 11, 12, 30, 32
coagulation, 3
cobalt, 25
co-existence, 32
communities, 13, 14, 37, 49
community, 15, 50
competition, 31, 32
components, 3, 15, 32
composition, 3, 4, 18, 19, 20, 24, 28
compounds, 2, 3, 5, 7, 8, 10, 11, 12, 13, 14, 15, 17, 18, 19, 21, 22, 23, 24, 25, 27, 29, 30, 31, 34, 36, 39, 44, 50
concentration, 3, 5, 7, 11, 22, 24, 25, 26, 28, 31, 32
control, 48
conversion, 35, 36, 45
copper, 25, 28, 49
costs, 2, 4, 27
cracking, 20, 21
crude oil, 7, 8, 9, 20, 21, 35, 48, 49
crystallization, 27
culture, 3, 4, 15, 22, 25, 27, 32
cycles, 27
cysteine, 14, 15
cytochrome, 12

D

dairy, 2, 8, 14, 15, 18, 19, 22, 43, 49

dairy industry, 2, 19, 49
decane, 9
decomposition, 9, 10, 15, 30, 31, 32, 39
definition, 17
degradation, 2, 43, 45, 46, 47, 48, 49
degrading, 43
demobilization, 28
denitrifying, 29
density, 25
deposits, 3, 4, 7, 8, 22, 24, 25, 26, 27, 28
derivatives, 43
desorption, 27
diesel, 20
diffusion, 15
digestion, 47
dissolved oxygen, 11
distillation, 20, 21
distribution, 7
DNA, 13
donor, 12, 13, 31
donors, 7, 10, 13, 20
drainage, 45
duration, 5, 32

E

earth, 20
ecological, 3
ecology, 45
effluent, 35, 36, 46
effluents, xi, 1, 2, 26, 28
electromagnetic, 20
electron, 7, 11, 12, 13, 17, 20, 31, 46
electrons, 8, 10, 34
electroplating, 26
emission, xi, 1, 26
emulsification, 20
emulsions, 20, 21
energy, 8, 10, 11, 12, 23, 33, 34
environment, xi, 1, 2, 3, 4, 7, 8, 11, 12, 13, 14, 18, 25, 26, 28, 30, 33, 34, 35, 36, 39, 48
enzymatic, 28
enzymes, 9, 20, 22, 30, 31
epoxy, 20
epoxy resins, 20

equilibrium, 2
erosion, 26
esters, 20
ethanol, 8, 10, 12, 13, 30, 32
ethylbenzene, 9, 46
ethylene, 20
exploitation, 12, 26
extraction, 3, 26

F

family, 34
farming, 22
farms, xi, 1
fat, 19
fats, 18
fatty acids, 10, 20, 21, 43, 45
fermentation, 9, 28, 29, 30, 31, 32
fertilizer, 26
fertilizers, 2, 21, 22, 26
fiber, 47
fillers, 24
film, 46
films, 21
filtration, 36
flotation, 48
flow, 3, 17, 24
fluid, 27
flushing, 18, 20, 21
food, xi, 1, 18, 29
food industry, 18
fractionation, 49
fresh water, 7
fuel, 20, 31
fumaric, 10, 13
fungi, 24

G

gas, 7, 20, 35
gaseous waste, 26
gases, xi, 1, 15
geothermal, 7
glucose, 8

glycol, 20
grains, 23, 24
granules, 24
groundwater, 8
groups, 10, 13, 15, 25, 27, 28, 29, 31, 32, 33, 34, 39
growth, 2, 3, 11, 12, 13, 15, 24, 30, 31, 35, 45
growth temperature, 11, 13

H

H_2, 12
hazards, 2
heat, 22
heating, 20
heavy metal, 4, 22, 25, 26, 27, 28, 32, 35, 36, 44, 46, 47
heavy metals, 4, 22, 25, 26, 27, 28, 32, 36, 44, 46, 47
herbicides, 18
hexane, 9
high temperature, 26
households, 17
human, xi, 1
humic acid, 47
hydro, 7, 18, 20, 21, 26, 45, 47, 48
hydrocarbon, 43
hydrocarbons, 20, 21, 48
hydrogen, 4, 5, 8, 10, 13, 18, 21, 30, 31, 32, 33, 34, 35, 36, 37, 46
hydrogen sulfide, 46
hydrolysis, 20, 29, 31
hydrometallurgy, 49
hydrothermal, 11
hydroxide, 21
hydroxides, 27

I

identification, 15
incubation, 3, 15
indicators, 14
indole, 9

industrial, xi, 1, 2, 3, 4, 17, 22, 25, 26, 28, 33, 39
industrial wastes, 3, 4, 22, 25, 39
industry, 2, 17, 18, 19, 20, 22, 26, 49
infrared, 35
infrared light, 35
inhibition, 44, 49
inhibitors, 12
inhibitory, 49
inorganic, 2, 3, 17, 21, 24, 25, 34, 39
inorganic salts, 21
ions, 10, 12, 22, 27, 28
iron, 10, 12, 15, 25, 29, 34, 44, 49
isolation, 14, 15

K

ketones, 45
killing, 11

L

lactic acid, 30
lactose, 18, 19, 20
lagoon, 43
lakes, 34
lanthanoids, 25
lanthanum, 25
leaching, 2, 48
leather, 20
limitation, 45
limitations, 27
lipids, 9
liquid phase, 2
liquidation, 49

M

magmatic, 26
malic, 8
management, 26, 27
manganese, 25, 28
meat, 18
media, 9, 13, 14, 15

membranes, 27
mercury, 25, 45
metabolic, 5, 43
metabolism, 3, 10, 28, 33, 36, 44
metabolites, 27, 45
metal ions, 27
metallurgy, 26, 28
metals, 4, 5, 22, 25, 26, 27, 28, 29, 35, 36, 48
methane, 30, 31
methanogenesis, 22, 32, 49
methanol, 8
methylation, 45
microbial, 43, 44, 46, 47, 48, 49
microbial communities, 49
microcosms, 14, 46, 50
microflora, 11
microorganism, 3, 5, 7, 14, 15, 24, 27, 29, 31, 32
microorganisms, 2, 3, 4, 8, 10, 12, 13, 15, 18, 23, 24, 27, 28, 29, 30, 31, 32, 33, 34, 39, 44, 45
milk, 2, 18, 19, 21
mineral oils, 21
mineralization, 5, 39, 45
minerals, 48
mines, 21
mining, xi, 1
mobility, 13, 25, 48
molasses, 22
molybdenum, 25
monomeric, 31
multiplication, 14

N

Na2SO4, 22
NaCl, 21
naphthalene, 15
natural, xi, 1, 2, 3, 4, 7, 13, 15, 17, 25, 26, 33, 34, 36, 37, 39, 43
natural environment, xi, 1, 2, 3, 4, 7, 13, 15, 25, 26, 33, 34, 36, 39
natural gas, 7
neutralization, 2, 26
Ni, 5

nickel, 25, 26, 28
Nielsen, 7, 8, 47
niobium, 25
nitrate, 47
nitrates, 13
nitrification, 2
nitrogen, xi, 1, 2, 19, 44
nitrogen compounds, xi, 1
nitrogen fixation, 44
normal, 20
norms, 3

O

octane, 9
oil, 2, 7, 8, 9, 13, 20, 21, 35, 46, 47, 48, 49
oil refineries, 2, 21
oils, 20, 21
optical, 35, 47
ores, 26, 35
organic, 2, 3, 5, 7, 8, 9, 10, 12, 13, 14, 15, 17, 18, 20, 22, 23, 24, 28, 29, 31, 32, 35, 36, 39, 49
organic compounds, 5, 7, 8, 9, 10, 12, 13, 14, 17, 18, 20, 22, 23, 28, 29, 31, 32, 35, 36
organic matter, 9, 15, 20, 32, 36, 49
oscillation, 5
oxidation, 8, 10, 12, 13, 21, 26, 30, 31, 33, 34, 36, 37, 44, 45, 47, 48, 49
oxide, 8
oxygen, 2, 11, 14, 15, 32
oxygenation, 15, 29
ozone, xi, 1

P

paints, 26
particles, 24, 26
passive, xi, 1
pathogenic, 4, 32
pesticide, 26
pesticides, 18, 26
petrochemical, 2, 7, 9, 13, 14, 15, 20, 39
petroleum, 20, 45, 47, 49, 50

petroleum products, 49
pH, 5, 11, 17, 18, 19, 21, 24, 28, 35, 47
pharmaceutical, 31
phenol, 9, 10, 14, 20, 21, 43, 47
phosphates, 5, 22, 36
phosphogypsum, 50
phosphorus, 2
phosphorylation, 10
photosynthetic, 46
phototrophic, 46
physiological, 15
physiology, 7
pipelines, 19
plants, xi, 1, 3, 4, 17, 18, 20, 21, 26, 35, 39, 44
plastic, 24, 26
plastics, 20, 31, 36
play, 22, 28
pollutant, 29
pollutants, xi, 1, 2, 3, 13, 14, 18, 20, 21, 22, 29, 31, 36
pollution, xi, 1, 2, 3, 24, 36
polycyclic aromatic hydrocarbon, 47
polymer, 26, 27
population, 27, 28
population density, 27
power, 22
power plant, 22
power plants, 22
precipitation, 8, 17, 26, 28, 32
pressure, 20, 26
printing, 26
production, xi, 1, 3, 17, 18, 20, 21, 22, 26, 31
profit, 10
prokaryotes, 49
propylene, 20
protection, 2, 39
protein, 12, 18, 19
proteins, 9, 10, 18, 19, 20
pulp, 20, 31
pyruvic, 8, 10, 13

R

radiation, 36
rain, 17
rainwater, 48
range, xi, 1, 2, 3, 11, 32, 39
rare earth, 3, 24
rare earth elements, 24
rare earths, 3
raw material, 20
reactive groups, 27
recovery, 25, 26, 35, 46
rectification, 20
redox, 11, 14, 15
REE, 25
refineries, 26
refining, 7, 20, 21, 47, 50
regulation, 28
relationship, 31, 32, 43
relationships, 5, 15, 31
reservoir, 3
reservoirs, 7, 8, 17, 47
resins, 21
resistance, 27, 46
respiration, 4, 7, 12, 44
rubber, 15, 20
runoff, 24

S

saline, 34
salts, 19, 20, 21, 22, 27
sand, 24
sanitation, 17
search, 3, 18
sedimentation, 20, 26
sediments, 8
sensitivity, 28
separation, 20, 24, 27
services, viii
sewage, xi, 1, 2, 3, 7, 13, 14, 15, 17, 18, 19, 20, 21, 22, 23, 24, 26, 27, 28, 29, 30, 31, 32, 33, 35, 36, 39, 44, 46
short-term, 28, 46
silver, 25
sites, 2, 48
slag, 24
sludge, 2, 4, 15, 31, 33, 47, 48
smelters, 22, 26

xylene, 9, 45

Y

yeast, 36

Z

zinc, 25, 26, 28, 49
zirconium, 25

sodium, 13, 14, 15, 21
sodium hydroxide, 21
soil, 2, 9, 26, 50
soils, 7, 9, 13, 49
solid phase, 24
solid waste, 1, 2, 4, 22, 27, 33, 36, 39
solubility, 25
solvent, 2
solvents, 3
sorption, 26, 27
species, 8, 9, 10, 11, 13, 30, 31, 34
spectrum, 13
speed, 3, 24
springs, 7
stability, 27
stabilization, 28
stages, 20, 22, 23, 25, 29, 31, 33, 36
steel, 7
stock, xi, 1
stoichiometry, 31
storage, xi, 1
strain, 9, 43, 46, 48
strains, 9, 15, 44
streams, 26, 44
substances, xi, 1, 4, 13, 27
substrates, 10, 31
sugar, 18, 22
sugar cane, 22
sulfate, 43, 44, 45, 46, 47, 48, 49
sulfur, 44, 45, 46
sulphate, 4, 7, 8, 10, 12, 13, 14, 18, 22, 23, 28, 29, 31, 34, 44, 45, 47, 48, 49, 50
sulphur, 3, 5, 7, 8, 10, 11, 12, 13, 17, 22, 25, 31, 32, 33, 34, 35, 36, 37, 44, 45, 46, 49
supply, 2, 3, 26, 34
surface water, 3
suspensions, 2, 21
symbiotic, 5

T

tanks, 2, 20, 24
tantalum, 25
taxonomic, 34
temperature, 5, 11, 17, 24, 26, 48

textile, 20, 26, 31
textile industry, 20
thallium, 25
tin, 25
TOC, 32
toluene, 9, 15, 43, 45, 48, 50
toxic, 3, 4, 5, 12, 18, 27, 28, 31, 32, 37, 39, 49
toxic metals, 28
toxicity, 28
transformation, 1, 26, 27, 33, 43
transformations, 30
transport, 10, 12, 19, 26
transportation, 26
treatment methods, 2, 17
tribes, 9, 11, 33
Turku, 50

U

uranium, 48

V

values, 18
vanadium, 25
vessels, 15, 18
vitamins, 19

W

waste treatment, 46
waste water, 23, 27, 43, 46
waste waters, 46, 47, 49, 50
wastes, 3, 4, 22, 25, 39
water, 2, 3, 13, 17, 20, 23, 26, 27, 28, 39, 44
water quality, 39
weathering, 26
whey, 8, 19, 50
wood, 22

X

xenobiotics, 18